Core 3

for Edexcel

CAMBRIDGE
UNIVERSITY PRESS

The School Mathematics Project

SMP AS/A2 Mathematics writing team Spencer Instone, John Ling, Paul Scruton, Susan Shilton, Heather West

SMP design and administration Melanie Bull, Pam Keetch, Nicky Lake, Cathy Syred, Ann White

The authors thank Sue Glover for the technical advice she gave when this AS/A2 project began and for her detailed editorial contribution to this book. The authors are also very grateful to those teachers who commented in detail on draft chapters.

CAMBRIDGE UNIVERSITY PRESS
Cambridge, New York, Melbourne, Madrid, Cape Town, Singapore, São Paulo

Cambridge University Press
The Edinburgh Building, Cambridge CB2 2RU, UK

www.cambridge.org
Information on this title: www.cambridge.org/9780521605373

© The School Mathematics Project 2005

First published 2005

Printed in the United Kingdom at the University Press, Cambridge

A catalogue record for this publication is available from the British Library

ISBN-13 978-0-521-60537-3 paperback
ISBN-10 0-521-60537-7 paperback

Typesetting and technical illustrations by The School Mathematics Project

The authors and publisher are grateful to London Qualifications Limited for permission to reproduce questions from past Edexcel examination papers. Individual questions are marked Edexcel. London Qualifications Limited accepts no responsibility whatsoever for the accuracy or method of working in the answers given.

Using this book

Each chapter begins with a **summary** of what the student is expected to learn.

The chapter then has sections lettered A, B, C, ... (see the contents overleaf). In most cases a section consists of development material, worked examples and an exercise.

The **development material** interweaves explanation with questions that involve the student in making sense of ideas and techniques. Development questions are labelled according to their section letter (A1, A2, ..., B1, B2, ...) and answers to them are provided.

D Some development questions are particularly suitable for discussion – either by the whole class or by smaller groups – because they have the potential to bring out a key issue or clarify a technique. Such **discussion questions** are marked with a bar, as here.

K **Key points** established in the development material are marked with a bar as here, so the student may readily refer to them during later work or revision. Each chapter's key points are also gathered together in a panel after the last lettered section.

The **worked examples** have been chosen to clarify ideas and techniques, and as models for students to follow in setting out their own work. Guidance for the student is in italic.

The **exercise** at the end of each lettered section is designed to consolidate the skills and understanding acquired earlier in the section. Unlike those in the development material, questions in the exercise are denoted by a number only.

Starred questions are more demanding.

After the lettered sections and the key points panel there may be a set of **mixed questions**, combining ideas from several sections in the chapter; these may also involve topics from earlier chapters.

Every chapter ends with a selection of **questions for self-assessment** ('Test yourself').

Included in the mixed questions and 'Test yourself' are **past Edexcel exam questions**, to give the student an idea of the style and standard that may be expected, and to build confidence.

Chapter 10 on **proof** gives students an opportunity to understand a variety of proofs and to prove some statements for themselves. Because it uses mathematical content that appears at various points in the Core 3 course, the chapter is placed at the end of the book; however it may be drawn on at appropriate points earlier in the course.

Contents

1 Rational expressions and division

In this chapter you will learn how to
- simplify rational expressions
- add, subtract, multiply and divide rational expressions

A Simplifying (answers p 154)

An expression that consists of one polynomial divided by another is called a **rational expression** or **algebraic fraction**.

A1 A function is defined by $f(n) = \dfrac{n+3}{n^2+4n+3}$ where n is an integer such that $n \geq 0$.

    (a) Evaluate each of these in its simplest fractional form.

        (i) $f(1)$ **(ii)** $f(4)$ **(iii)** $f(10)$

    (b) **(i)** Without calculating, what do you think is the value of $f(100)$ in its simplest form?

        (ii) Check your result.

    (c) **(i)** What do you think is the value of $f(k)$ in its simplest fractional form?

        (ii) Prove your result.

    (d) Show that $f(n) \leq \frac{1}{2}$ for all positive integer values of n.

A2 A function is defined by $f(n) = \dfrac{n^2+6n+5}{n^2+7n+10}$ where n is an integer such that $n \geq 0$.

    (a) Evaluate each of these in its simplest fractional form.

        (i) $f(0)$ **(ii)** $f(3)$ **(iii)** $f(20)$

    (b) **(i)** What do you think is the value of $f(k)$ in its simplest fractional form?

        (ii) Prove your result.

    (c) Hence show that the equation $f(n) = \frac{8}{11}$ has no integer solution.

A3 Prove that, when x is a multiple of 5, the value of the expression $\dfrac{5x+10}{x^2+2x}$ can be written as a unit fraction (with 1 as its numerator).

When simplifying rational expressions, it is usually beneficial to factorise whenever possible.

Example 1

Simplify $\dfrac{3x-12}{x^2-4x}$.

Solution

Factorise.
$$\frac{3x-12}{x^2-4x} = \frac{3(x-4)}{x(x-4)}$$

Divide numerator and denominator by $(x-4)$.
$$= \frac{3}{x}$$

Example 2

Simplify $\dfrac{n^2 + 2n - 15}{2n + 1} \times \dfrac{1}{n + 5}$.

Solution

$$\frac{n^2 + 2n - 15}{2n + 1} \times \frac{1}{n + 5} = \frac{n^2 + 2n - 15}{(2n + 1)(n + 5)}$$

Factorise.

$$= \frac{(n + 5)(n - 3)}{(2n + 1)(n + 5)}$$

Divide numerator and denominator by $(n + 5)$.

$$= \frac{n - 3}{2n - 1}$$

You may need to revise dividing by a fraction.

For example, $\dfrac{3}{4} \div \dfrac{5}{2} = \dfrac{\frac{3}{4}}{\frac{5}{2}} = \dfrac{\frac{3}{4} \times \frac{2}{5}}{\frac{5}{2} \times \frac{2}{5}} = \dfrac{\frac{3}{4} \times \frac{2}{5}}{1} = \dfrac{3}{4} \times \dfrac{2}{5}$.

 In general, dividing by a fraction is equivalent to multiplying by its reciprocal. This rule applies to all rational expressions: $\dfrac{a}{b} \div \dfrac{c}{d} = \dfrac{a}{b} \times \dfrac{d}{c}$.

Example 3

Simplify $\dfrac{4n + 4}{n^2 - 9} \div \dfrac{8}{2n^2 + 5n - 3}$.

Solution

Use the rule for dividing.

$$\frac{4n + 4}{n^2 - 9} \div \frac{8}{2n^2 + 5n - 3} = \frac{4n + 4}{n^2 - 9} \times \frac{2n^2 + 5n - 3}{8}$$

Factorise.

$$= \frac{4(n + 1)}{(n + 3)(n - 3)} \times \frac{(2n - 1)(n + 3)}{8}$$

Divide numerator and denominator by 4 and $(n + 3)$.

$$= \frac{(n + 1)(2n - 1)}{2(n - 3)}$$

Exercise A (answers p 154)

1 Simplify each of these.

(a) $\dfrac{x^2 + 6x}{2x + 12}$ (b) $\dfrac{x^2 - 3x}{x^2 + x}$ (c) $\dfrac{3x + 6}{2 + x}$ (d) $\dfrac{x - 3}{5x - 15}$ (e) $\dfrac{6n^2 + 4n}{4n^2 + 2n}$

2 (a) Show that $\dfrac{14x - 4x^2}{2x - 7}$ is equivalent to $-2x$.

(b) Simplify each of these.

(i) $\dfrac{15n - 10}{2 - 3n}$ (ii) $\dfrac{n^2 - 7n}{7 - n}$

3 Simplify each of these.

(a) $\dfrac{x^2 + 5x + 6}{x + 2}$ (b) $\dfrac{x^2 + 2x - 3}{2x + 6}$ (c) $\dfrac{x - 4}{x^2 - 2x - 8}$ (d) $\dfrac{x^2 + 5x}{x^2 - 25}$ (e) $\dfrac{3n^2 - n}{9n^2 - 1}$

4 A function is defined by $g(n) = \dfrac{2n + 18}{n^2 + 10n + 9}$ where n is a positive integer.

(a) Evaluate $g(3)$ and $g(4)$ in their simplest fractional form.

(b) When n is odd, prove that in its simplest form $g(n)$ is a unit fraction.

(c) Find the value of n such that $g(n) = \frac{2}{9}$.

5 Simplify each of these.

(a) $\dfrac{n^2 + 9n + 20}{n^2 + 11n + 30}$ (b) $\dfrac{n^2 + 5n - 14}{n^2 - 7n + 10}$ (c) $\dfrac{2n^2 - 10n - 12}{n^2 + 6n + 5}$ (d) $\dfrac{3n^2 - 16n + 5}{n^2 - 3n - 10}$

(e) $\dfrac{2n^2 + n - 3}{2n^2 + 7n + 6}$ (f) $\dfrac{2n^2 + 7n - 4}{3n^2 - 48}$ (g) $\dfrac{n^2 + n}{n^3 - n}$ (h) $\dfrac{n^2 + 5n - 24}{n^3 - 9n}$

6 Simplify each of these.

(a) $\dfrac{2x - 10}{x + 1} \times \dfrac{3x + 3}{x - 5}$ (b) $\dfrac{x^2 - 16}{x - 4} \times \dfrac{2x}{x + 4}$

(c) $\dfrac{x^2 + 7x + 12}{x + 2} \times \dfrac{5}{x + 3}$ (d) $\dfrac{6x - 12}{3x + 15} \times \dfrac{x^2 + 10x + 25}{x^2 - 4x + 4}$

(e) $\dfrac{x^2 + 2x - 3}{x^2 + 6x + 9} \times \dfrac{x^2 + 10x + 21}{x^2 - 2x + 1}$ (f) $\dfrac{4x^2 - 9}{2x^2 + 13x + 15} \times \dfrac{x^2 - 25}{2x^2 - 5x + 3}$

7 Simplify each of these.

(a) $\dfrac{2}{x^2 + x} \div \dfrac{4}{5x + 5}$ (b) $\dfrac{x^2 - 1}{2x + 3} \div \dfrac{x^2 - x}{4x + 6}$

(c) $\dfrac{2x^2 + x - 15}{x} \div \dfrac{2x^2 - 13x + 20}{x^2 - 4x}$ (d) $\dfrac{9x^2 - 4}{2x^2 - 4x - 70} \div \dfrac{6x^2 - 7x + 2}{2x^2 + 9x - 5}$

8 Show that $\dfrac{1}{x} \div y$ is equivalent to the single fraction $\dfrac{1}{xy}$.

9 Simplify each of these.

(a) $\dfrac{1}{x} \div \dfrac{1}{y}$ (b) $\dfrac{1}{x} \div 3$ (c) $6x \div \dfrac{3}{y}$ (d) $\dfrac{4}{x^2} \div \dfrac{1}{2x}$ (e) $\frac{1}{5} \div 10x$

10 Show that $\dfrac{6}{\left(\dfrac{2}{x - 1}\right)} = 3(x - 1)$.

11 Functions are defined by $f(x) = \dfrac{6x+8}{3x^2+10x+8}$ and $g(x) = \dfrac{2}{x^2+4x+4}$ where x is a positive integer.

   (a) **(i)** Evaluate $f(2)$ and $g(2)$ in their simplest fractional form.

      (ii) Find $\dfrac{f(2)}{g(2)}$ in its simplest form.

   (b) Prove that $\dfrac{f(x)}{g(x)}$ is always an integer.

B Adding and subtracting

Fractions can easily be added or subtracted if they are written with the same denominator. The lowest common multiple of two denominators is called the **lowest common denominator** and is usually the simplest to use.

For example, $\frac{3}{4} - \frac{2}{3} = \frac{9}{12} - \frac{8}{12} = \frac{9-8}{12} = \frac{1}{12}$.

Algebraic fractions can be dealt with in the same way.

Example 4

Express $\dfrac{1}{x-2} - \dfrac{3}{5x}$ as a single fraction in its simplest form.

Solution

A suitable denominator is $5x(x-2)$. $\dfrac{1}{x-2} - \dfrac{3}{5x} = \dfrac{5x}{5x(x-2)} - \dfrac{3(x-2)}{5x(x-2)}$

$$= \dfrac{5x - 3(x-2)}{5x(x-2)}$$

Expand the brackets in the numerator. $= \dfrac{5x - 3x + 6}{5x(x-2)}$

Simplify. $= \dfrac{2x+6}{5x(x-2)}$

Factorise if possible. $= \dfrac{2(x+3)}{5x(x-2)}$

Example 5

Express $4 + \dfrac{3}{2x+1}$ as a single fraction in its simplest form.

Solution

Write 4 as a fraction with a denominator of $2x+1$. $4 + \dfrac{3}{2x+1} = \dfrac{4(2x+1)}{2x+1} + \dfrac{3}{2x+1}$

$$= \dfrac{4(2x+1)+3}{2x+1}$$

Expand the brackets and simplify. $= \dfrac{8x+7}{2x+1}$

Example 6

Express $\dfrac{x+1}{2x-1} + \dfrac{2}{x-5}$ as a single fraction in its simplest form.

Solution

A suitable denominator is $(2x-1)(x-5)$.

$$\dfrac{x+1}{2x-1} + \dfrac{2}{x-5} = \dfrac{(x+1)(x-5)+2(2x-1)}{(2x-1)(x-5)}$$

Expand the brackets on the numerator.

$$= \dfrac{x^2-4x-5+4x-2}{(2x-1)(x-5)}$$

Simplify.

$$= \dfrac{x^2-7}{(2x-1)(x-5)}$$

Example 7

Express $\dfrac{x}{(x+2)(x+3)} - \dfrac{6}{(x+2)(x-1)}$ as a single fraction in its simplest form.

Solution

A suitable denominator is $(x-1)(x+2)(x+3)$.

$$\dfrac{x}{(x+2)(x+3)} - \dfrac{6}{(x+2)(x-1)} = \dfrac{x(x-1)-6(x+3)}{(x-1)(x+2)(x+3)}$$

Expand the brackets in the numerator.

$$= \dfrac{x^2-x-6x-18}{(x-1)(x+2)(x+3)}$$

Simplify.

$$= \dfrac{x^2-7x-18}{(x-1)(x+2)(x+3)}$$

Factorise.

$$= \dfrac{(x-9)(x+2)}{(x-1)(x+2)(x+3)}$$

Cancel.

$$= \dfrac{x-9}{(x-1)(x+3)}$$

Exercise B (answers p 155)

1 Express each of these as a single fraction in its simplest form.

(a) $\dfrac{1}{x} + \dfrac{1}{2x+3}$ (b) $\dfrac{1}{x} - \dfrac{1}{x+5}$ (c) $\dfrac{3}{x-1} + \dfrac{4}{3x}$ (d) $\dfrac{2}{x-1} - \dfrac{2}{x}$

2 Express each of these as a single fraction in its simplest form.

(a) $x + \dfrac{5}{x}$ (b) $\dfrac{9}{2x+1} + 3$ (c) $\dfrac{x}{x-4} - 5$ (d) $\dfrac{x}{3x-2} - x$

3 Express each of these as a single fraction.

(a) $\dfrac{1}{a} + \dfrac{1}{b}$ (b) $\dfrac{2}{x} - \dfrac{1}{y}$ (c) $a + \dfrac{3}{b}$ (d) $\dfrac{a}{c} - b$ (e) $\dfrac{1}{2a} - \dfrac{1}{3b}$

4 Express $\dfrac{a}{b} - \dfrac{a}{b+1}$ as a single fraction in its simplest form.

5 Express each of these as a single fraction in its simplest form.

(a) $\dfrac{1}{x+1} + \dfrac{1}{x-1}$
(b) $\dfrac{2}{x+3} - \dfrac{1}{3x+1}$
(c) $\dfrac{x-1}{x-2} + \dfrac{6}{x-5}$

(d) $\dfrac{x}{x+4} - \dfrac{x-5}{x-1}$
(e) $\dfrac{x+1}{x+3} + \dfrac{x-4}{2x-1}$
(f) $\dfrac{x-1}{x-7} - \dfrac{x+7}{x+1}$

6 Express each of these as a single fraction in its simplest form.

(a) $\dfrac{6}{x} + \dfrac{3}{x(x-4)}$
(b) $\dfrac{2x}{(x+3)(x-1)} - \dfrac{1}{(x-1)}$

(c) $\dfrac{x+5}{(x+7)(x+1)} + \dfrac{x-1}{(x+1)(x+4)}$
(d) $\dfrac{1}{(x-7)(2x-1)} - \dfrac{1}{(2x-1)(x+1)}$

(e) $\dfrac{1}{ab} + \dfrac{1}{bc}$
(f) $\dfrac{z}{3xy} - \dfrac{x}{4yz}$

7 (a) Factorise the denominators in the sum $\dfrac{4}{2x^2 - x - 1} + \dfrac{12}{2x^2 + 7x + 3}$.

(b) Show that this sum is equivalent to $\dfrac{16x}{(x-1)(2x+1)(x+3)}$.

8 Express each of these as a single fraction in its simplest form.

(a) $\dfrac{6}{x+5} + \dfrac{8}{2x^2 + 9x - 5}$
(b) $\dfrac{3}{x^2 + 2x} + \dfrac{1}{x^2 + 6x + 8}$

(c) $\dfrac{x+11}{x^2 - 9} - \dfrac{4}{x^2 + 3x}$
(d) $\dfrac{3x+5}{x^2 + 3x + 2} - \dfrac{2x+1}{x^2 + x - 2}$

9 Express $1 - \dfrac{2}{x+5} + \dfrac{x-15}{x^2 - 25}$ as a single fraction in its simplest form.

10 A function is defined by

$$f(x) = \frac{x+2}{x+3} + \frac{2x+3}{x^2 + 3x},$$

where x is positive.

(a) Evaluate $f(5)$ as a single fraction in its simplest form.

(b) Prove that $f(x) > 1$ for all positive values of x.

11 (a) Express $\dfrac{1}{x} + \dfrac{1}{2}$ as a single fraction.

(b) Hence write the expression $\dfrac{1}{\frac{1}{x} + \frac{1}{2}}$ as a single fraction.

12 Show that $\dfrac{1}{\frac{3}{x-4} - 1} = \dfrac{x-4}{7-x}$.

C Extension: Leibniz's harmonic triangle (answers p 155)

This section provides an opportunity to apply the techniques of sections A and B to some fraction patterns. It also provides valuable practice in proving statements. The method introduced to add a series by writing each term as a difference is not part of the content for Core 3 or 4.

Gottfried Leibniz (1646–1716) was a German philosopher and mathematician who is best known for his work on calculus. The distinguished Dutch physicist and mathematician Christian Huygens (1629–1693) challenged Leibniz to calculate the infinite sum of the reciprocals of the triangle numbers:

$$\tfrac{1}{1} + \tfrac{1}{3} + \tfrac{1}{6} + \tfrac{1}{10} + \dots$$

D

C1 Show that the nth term of this series is $\dfrac{2}{n(n+1)}$.

(You need to know that the nth triangle number is $\tfrac{1}{2}n(n + 1)$.)

C2 (a) Show that each term can be written as the difference $\dfrac{2}{n} - \dfrac{2}{n+1}$.

(b) Hence, show that the sum of the first n terms can be written:

$$\left(\tfrac{2}{1} - \tfrac{2}{2}\right) + \left(\tfrac{2}{2} - \tfrac{2}{3}\right) + \left(\tfrac{2}{3} - \tfrac{2}{4}\right) + \left(\tfrac{2}{4} - \tfrac{2}{5}\right) + \left(\tfrac{2}{5} - \tfrac{2}{6}\right) + \dots + \left(\dfrac{2}{n} - \dfrac{2}{n+1}\right)$$

(c) Hence find a formula for the sum of the reciprocals of the first n triangle numbers. Write your formula as a single fraction.

(d) Use your formula to find the sum of the reciprocals of the first five triangle numbers. Check your result by adding the appropriate fractions.

C3 Now, think about the sum to infinity of the reciprocals of the triangle numbers. Show that, as n gets larger, the sum gets closer and closer to 2.

Exercise C (answers p 156)

In the course of his work on summing infinite series, Leibniz devised a triangle which he called the **harmonic triangle**.

Part of this triangle is

$$
\begin{array}{ccccccccc}
 & & & & 1 & & & & \\
 & & & \tfrac{1}{2} & & \tfrac{1}{2} & & & \\
 & & \tfrac{1}{3} & & \tfrac{1}{6} & & \tfrac{1}{3} & & \\
 & \tfrac{1}{4} & & \tfrac{1}{12} & & \tfrac{1}{12} & & \tfrac{1}{4} & \\
 \tfrac{1}{5} & & \tfrac{1}{20} & & \tfrac{1}{30} & & \tfrac{1}{20} & & \tfrac{1}{5}
\end{array}
$$

- The fractions on each edge form a sequence of unit fractions where the denominators increase by 1 each time.

- Each fraction in the triangle is the sum of the two fractions below it.

1 Verify that the sum of $\tfrac{1}{20}$ and $\tfrac{1}{30}$ is $\tfrac{1}{12}$.

2 In their simplest form, find the fractions in the next row of the harmonic triangle.

Consider the fractions in the diagonals of the triangle.

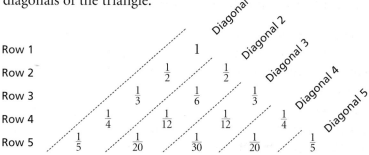

Row 1

Row 2

Row 3

Row 4

Row 5

3 What fraction appears in row 10 and diagonal 2?

4 (a) Show that in diagonal 1 the kth fraction and its successor can be written as $\dfrac{1}{k}$ and $\dfrac{1}{k+1}$.

(b) Show that the sum of the first n fractions in diagonal 2 can be written as
$$\left(\tfrac{1}{1} - \tfrac{1}{2}\right) + \left(\tfrac{1}{2} - \tfrac{1}{3}\right) + \left(\tfrac{1}{3} - \tfrac{1}{4}\right) + \left(\tfrac{1}{4} - \tfrac{1}{5}\right) + \left(\tfrac{1}{5} - \tfrac{1}{6}\right) + \dots + \left(\tfrac{1}{n} - \tfrac{1}{n+1}\right)$$

(c) Hence find a formula for the sum of the first n fractions in diagonal 2. Write your formula as a single fraction.

(d) Use your formula to find the sum of the first four fractions in diagonal 2. Check your result by adding the appropriate fractions.

(e) What will happen to the sum of the first n fractions in diagonal 2 as n gets larger and larger? Justify your answer.

(f) Prove that the kth fraction in diagonal 2 can be written as $\dfrac{1}{k(k+1)}$.

(g) (i) Show that the fraction $\tfrac{1}{420}$ appears in diagonal 2.

(ii) In which row is $\tfrac{1}{420}$?

5 (a) Show that in diagonal 2 the kth fraction and its successor can be written as $\dfrac{1}{k(k+1)}$ and $\dfrac{1}{(k+1)(k+2)}$.

(b) Hence find a formula for the sum of the first n fractions in diagonal 3.

(c) What will happen to the sum of the first n fractions in diagonal 3 as n gets larger and larger? Justify your answer.

(d) (i) Prove that the kth fraction in diagonal 3 can be written as $\dfrac{2}{k(k+1)(k+2)}$.

(ii) What is the 10th fraction in diagonal 3?

*6 Investigate the other diagonals in the harmonic triangle.
Can you find an expression for the nth fraction in diagonal m?
Can you find an expression for the sum of the first n fractions in diagonal m?
What happens to the sum of the first n fractions in diagonal m as n gets larger?

*7 Prove that the sum of the reciprocals of any pair of consecutive triangle numbers T_n and T_{n+1} is $\dfrac{4}{T_n + T_{n+1} - 1}$.

D Extension: the harmonic mean (answers p 157)

This section provides an opportunity to apply the techniques of sections A and B to a new type of average, the harmonic mean. It includes some challenging work on proving statements. The harmonic mean itself is not part of the content for Core 3 or 4.

You will be familiar with the arithmetic mean of two numbers (half of the sum) and possibly their geometric mean (the square root of the product). The harmonic mean was probably so called because it can be used to produce a set of harmonious notes in music.

One of the earliest mentions of it is in a surviving fragment of the work of Archytas of Tarentum (*circa* 350 BCE) who was a contemporary of Plato. He wrote 'There are three means in music: one is arithmetic, the second is geometric, and the third is the subcontrary, which they call harmonic.'

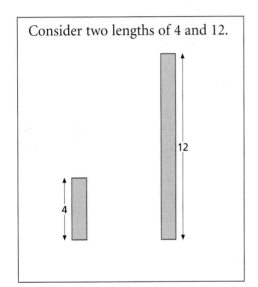

Consider two lengths of 4 and 12.

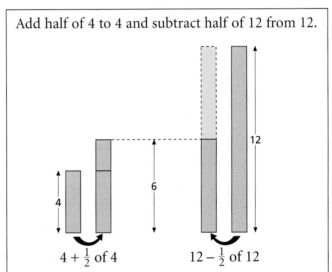

Add half of 4 to 4 and subtract half of 12 from 12.

$4 + \frac{1}{2}$ of 4 $12 - \frac{1}{2}$ of 12

$4 + \left(\frac{1}{2} \text{ of } 4\right) = 6$ and $12 - \left(\frac{1}{2} \text{ of } 12\right) = 6$, so we say that the harmonic mean of 4 and 12 is 6.

For any two numbers, if you find a fraction $\frac{p}{q}$ so that

smaller number $+ \left(\frac{p}{q} \text{ of smaller number}\right)$ is the same as *larger number* $- \left(\frac{p}{q} \text{ of larger number}\right)$

then the value of these two expressions is the **harmonic mean** of the two numbers.

D1 (a) Write down   (i) $6 + \left(\frac{2}{3} \text{ of } 6\right)$ (ii) $30 - \left(\frac{2}{3} \text{ of } 30\right)$

  (b) Hence write down the harmonic mean of 6 and 30.

D2 Try various fractions until you find the harmonic mean of

  (a) 6 and 18   (b) 10 and 15   (c) 4 and 28

D3 (a) If k is a fraction so that $a + ka = b - kb$, find an expression for k in terms of a and b.

  (b) Hence show that the harmonic mean of a and b can be expressed as $\dfrac{2ab}{a+b}$.

  (c) Use this rule to find the harmonic mean of 10 and 90.

Key points

- When working with rational expressions
  - factorise all expressions where possible
  - cancel any factors common to the numerator and denominator   (pp 6–10)

- To divide by a rational expression you can multiply by its reciprocal.   (p 7)

- To add or subtract rational expressions, write each with the same denominator.   (pp 9–10)

- An algebraic fraction is proper if the degree of the polynomial that is the numerator is less than the degree of the polynomial that is the denominator.   (p 16)

- When a polynomial $p(x)$ is divided by $d(x)$, a polynomial that is not a factor of and is of a degree less than or equal to $p(x)$, then we can write
$$\frac{p(x)}{d(x)} = q(x) + \frac{R(x)}{d(x)}, \text{ where } \frac{R(x)}{d(x)} \text{ is a proper fraction.}$$ (pp 16–17)

Test yourself (answers p 159)

1 Express as a single fraction in its simplest form $\dfrac{x^2 - 3x + 2}{x^2 - 4} \div \dfrac{x^2 - 2x + 1}{5x^2 + 10x}$.

2 Show that $\dfrac{x^2 + 5x + 4}{x^2 + 4x} = 1 + \dfrac{1}{x}$.

3 Express each of these as a single fraction, in its simplest form where appropriate.

(a) $\dfrac{2}{x} - \dfrac{1}{x+3}$ (b) $\dfrac{2}{x+4} + \dfrac{1}{x-2}$ (c) $\dfrac{x}{y} + \dfrac{y}{x}$

4 Show that $\dfrac{6}{(y-3)(y-1)} - \dfrac{3}{y-3} = \dfrac{3}{1-y}$.

5 Express $\dfrac{y+3}{(y+1)(y+2)} - \dfrac{y+1}{(y+2)(y+3)}$ as a single fraction in its simplest form.   Edexcel

6 Express $\dfrac{2x}{(x+1)(x+5)} + \dfrac{x+30}{x^2-25}$ as a single fraction in its simplest form.

7 Express $\dfrac{12}{x^2-3x} + \dfrac{x-19}{x^2-2x-3}$ as a single fraction in its simplest form.

8 Find the values of the constants A, B and C so that
$$\frac{3x^2 + 14x + 1}{x+5} = Ax + B + \frac{C}{x+5}$$

9 Show that $2x + \dfrac{1}{x-1} - \dfrac{5}{x^2+3x-4} = \dfrac{2x^2+8x+1}{x+4}$.

2 Functions

In this chapter you will learn
- what is meant by a function, including one–one and many–one functions
- about the domain and range of a function
- how to find a composite function
- how to find an inverse function and draw its graph

A What is a function? (answers p 160)

The concept of a **function** began its development in the 18th century and is now fundamental to almost every branch of mathematics. A function is essentially a rule or process that generates exactly one output for every given input.

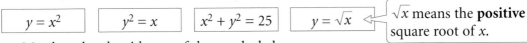

An example of a function is one where the inputs are people and the outputs are their favourite colours. So an input could be Jane Jones and the output would be her favourite colour, say, red.

In mathematics, we often deal with functions where both the inputs and outputs are numbers.

For a rule such as $y = x + 2$, we can think of the values of x as the inputs and the values of y as the outputs. For example, an input of $x = 5$ gives an output of $y = 5 + 2 = 7$.

Not all rules generate exactly one output for each given input: not all rules give functions.

A1 Below are four rules that connect x and y.

$$y = x^2 \qquad y^2 = x \qquad x^2 + y^2 = 25 \qquad y = \sqrt{x}$$

$\sqrt{x}$ means the **positive** square root of x.

(a) Match each rule with one of the graphs below.

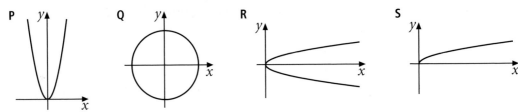

(b) For each rule, find the corresponding value or values of y when $x = 4$.

$y = x^2 + 1$ is an example of a rule for a function as each input value of x generates **exactly one** corresponding output value of y.

For example, when $x = 3$ then $y = 10$.

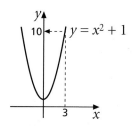

However, $x^2 + y^2 = 25$ is not a rule for a function as each input value of x between -5 and 5 generates **two** corresponding values of y.

For example, when $x = 3$ then $y = 4$ and $y = -4$.

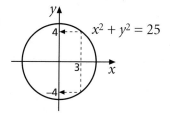

A2 Which of these are not rules for functions?

| $y = x^2 - 3$ | $y^2 - x^2 = 0$ | $y = \sqrt{x+3}$ | $y^3 = x$ | $y^4 = x$ |

A3 For the rule $y = \sqrt{x}$, can you find the value of y when $x = -5$?

When defining a function you need to specify the set of input values to be used. This set of input values is called the **domain** of the function.

For example, the rule $y = \sqrt{x}$ generates one value of y for each input value of x that is positive or 0. We cannot find the square root of a negative number so we can use the rule to define a function if we use non-negative input values. So a suitable domain is the set of values $x \geq 0$.

A definition of a function consists of two parts.

The **rule**   This tells you how values of the function are calculated.

The **domain**   This tells you the set of values to which the rule is applied.

We can use letters to stand for functions, the most usual ones being f, g and h. For example, if we call the square root function f, we can write it as

$$f(x) = \sqrt{x}, \ x \geq 0$$

where the rule is $f(x) = \sqrt{x}$ and the domain is the set of values $x \geq 0$.

(We could write the same function as $f(y) = \sqrt{y}, \ y \geq 0$: using y to stand for the input values does not change the rule or the domain.)

Sometimes, the context in which a rule is applied can restrict the domain.

For example, the rule $g(x) = x^3$ can be applied to any real number x and so the domain of g could be the complete set of real numbers (the complete set of rational and irrational numbers).

However, if we use the rule $g(x) = x^3$ to determine the volume of a cube of side x, it would be inappropriate to include negative values in the domain.

A4 A circle is cut out of a square of metal as shown.
The centre of the circle is at the centre of the square.

The radius of the circle is r.
The area of metal remaining is A(r).

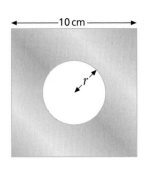

(a) Which of the following is the rule for the function A?

$A(r) = 10 - \pi r^2$	$A(r) = \pi r^2 - 100$
$A(r) = 100 - \pi r^2$	$A(r) = \pi r^2 - 10$

(b) Find the value of A(3), correct to two decimal places.

(c) Which of the following do you think is an appropriate domain for the function A?

$r \le 100$	$r \le 5$	$0 \le r \le 10$	$0 \le r \le 100$	$0 \le r \le 5$	$r \le 10$

To sketch a graph of a function, you need to exclude input values that are not in its domain.

Example 1

Sketch the graph of $y = f(x)$ where $f(x) = x^2 + 1, \; x \ge -2$.

Solution

The graph of the rule $y = x^2 + 1$ for all real values of x is as shown.

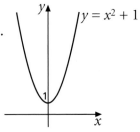

The domain of f is $x \ge -2$.
$f(-2) = (-2)^2 + 1 = 5$ so
the graph of $y = f(x)$ is as shown.

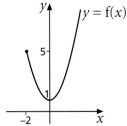

The point $(-2, 5)$ is included in the graph, so is shown by a solid circle.

An open circle can be used to show that an end-point is not included in a graph.

A useful shorthand for the set of real numbers is $\mathbb{R}$.

The statement '$x \in \mathbb{R}$' means 'x belongs to the set of real numbers'.

Sometimes we want the domain of a function to be all real numbers except a particular value.

For example, $f(t) = \dfrac{1}{t-2}$ is not defined for $t = 2$ (as division by 0 is not possible)

so a suitable domain for f is $t \in \mathbb{R}, t \ne 2$ (the set of all real numbers except 2).

A5 Sketch the graph of $y = f(x)$ for each function below.

    (a) $f(x) = x^2 - 1, \ x \leq 3$ **(b)** $f(x) = 2^x, \ x \in \mathbb{R}$

    (c) $f(x) = 3x + 1, \ x > -1$ **(d)** $f(x) = \dfrac{2}{x}, \ x \in \mathbb{R}, \ x \neq 0$

There is an alternative notation for the rule of a function.
For example, rather than writing $f(x) = x^2$ we can write

    $f: x \mapsto x^2$ (Read 'f: $x \mapsto x^2$' as 'f, such that x maps to x^2')

This notation suggests the idea of the input x being 'converted' to the output x^2.
You can also use an ordinary arrow ($\rightarrow$) for this purpose.

For example, the input 3 is converted to the input 3^2 or 9.
We could write either $f(3) = 9$ or $f: 3 \mapsto 9$.

A6 Sketch the graph of $y = f(x)$ where $f: x \mapsto \sqrt{x+2}, \ x \geq -2$.

Exercise A (answers p 160)

1 Match each function to its sketch graph.

    (a) $f(x) = x^2, \ x \geq 1$ **(b)** $f(x) = x^2, \ x \geq -2$ **(c)** $f(x) = x + 2, \ x \in \mathbb{R}$

    (d) $f(x) = x + 2, \ x \leq 0$ **(e)** $f(x) = \sqrt{x}, \ x \geq 1$ **(f)** $f(x) = \sqrt{x-1}, \ x \geq 1$

    (g) $f(x) = \dfrac{1}{x}, \ x \in \mathbb{R}, \ x \neq 0$ **(h)** $f(x) = \dfrac{1}{x}, \ x > 1$

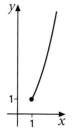
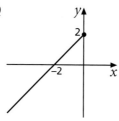
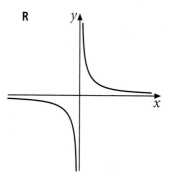
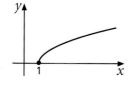
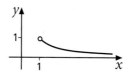
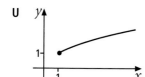
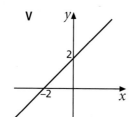
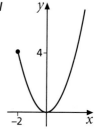

2 Sketch the graph of $y = f(x)$ for each of these functions.

(a) $f(x) = x - 2, \ x \leq 1$ (b) $f(x) = x^2 - 3, \ x > -2$

(c) $f: x \mapsto 9 - x^2, \ x \in \mathbb{R}$ (d) $f: x \mapsto \sqrt{x} + 2, \ x \geq 0$

(e) $f(x) = x - 3, \ -1 \leq x \leq 10$ (f) $f(x) = x^2, \ -2 < x < 4$

3 A rule for a function is $f(c) = \dfrac{1}{c + 3}$.

(a) Evaluate   (i) $f(2)$ (ii) $f(-2)$ (iii) $f(0)$ (iv) $f(-5)$

(b) Which value of c cannot be included in the domain of f?

B Many–one and one–one functions (answers p 161)

B1 (a) For the function $f(x) = 2x + 7, \ x \in \mathbb{R}$, solve the equation $f(x) = 25$.

(b) For the function $g(x) = x^2, \ x \in \mathbb{R}$, solve the equation $g(x) = 4$.

(c) For the function $h(x) = x^3, \ x \in \mathbb{R}$, solve the equation $h(x) = 27$.

K A **many–one function** is a function where there are two or more different inputs that generate the same output.

For example, for the function $f(x) = x^2$

we have $f(2) = 2^2 = 4$
and $f(-2) = (-2)^2 = 4$

Two different inputs (-2 and 2) generate the same output (4) so the function $f(x) = x^2$ is many–one.

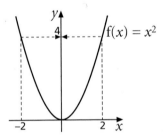

K Any function which is not many–one is a **one–one function**. Each output can be generated by only one input.

B2 A function g is defined for all real values of x by $g(x) = x^3 - 3x + 1$. A sketch of $y = g(x)$ is shown. Explain how the sketch shows that g is many–one.

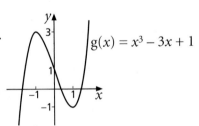

B3 A function is defined by $f(x) = x^3 - x, \ x \in \mathbb{R}$.

(a) Evaluate $f(1)$, $f(0)$ and $f(-1)$.

(b) Is f a one–one function?

B4 Classify each function as one–one or many–one.

(a) $f(x) = x^2 - 2, \ x \in \mathbb{R}$ (b) $f(n) = n^3, \ n \in \mathbb{R}$

(c) $f: t \mapsto 3t + 5, \ t \in \mathbb{R}$ (d) $f: x \mapsto x^4, \ x \in \mathbb{R}$

1 Two functions are defined as

$$f(x) = x^2 + 1, \ x < 1$$
$$g(x) = x^2 + 1, \ x > 1$$

(a) Draw a sketch graph for each function.

(b) Which one of these functions is one–one?

2 A function is defined by $f(t) = t^2 + 2t, \ t \geq -2$.

(a) Solve the equation $f(t) = 0$.

(b) Show that the equation $f(t) = 3$ has only one solution.

3 (a) Evaluate $\sin 0$ and $\sin \pi$.

(b) Show that the function $g(\theta) = \sin \theta$ with domain $0 \leq \theta \leq 2\pi$ is many–one.

(c) Solve the inequality $g(\theta) > \frac{1}{2}$.

4 A function is defined by $h: x \mapsto \cos x, \ 0 \leq x \leq \pi$.
Is the function h one–one or many–one? Explain how you decided.

5 A square is cut out of a square piece of metal as shown. The point O is the centre of both squares.

The length of one edge of the smaller square is x cm. The area of metal remaining is $A(x)$ cm^2.

(a) Show that the rule for the function A is

$$A(x) = 144 - x^2$$

(b) What is a suitable domain for the function A?

(c) (i) Sketch the graph of $y = A(x)$.

(ii) Classify the function A as one–one or many–one.

(d) Solve the equation $A(x) = 100$.

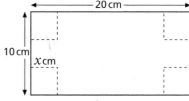

*6 A manufacturer has some sheets of card. Each sheet measures 20 cm by 10 cm.

A cuboid-shaped box is to be made from each sheet by cutting a square (x cm by x cm) from each corner and folding up.

The sheet of card is assumed to be of negligible thickness and rigid.

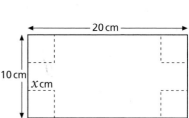

(a) Show that the volume of the box, $V(x)$ cm^3, is given by

$$V(x) = 4x^3 - 60x^2 + 200x$$

(b) What is a suitable domain for the function V?

(c) Show that the function V is many–one.

(d) The manufacturer wants to make boxes with a volume of 144 cm^3. What value of x should the manufacturer use?

C The range of a function

The **range** of a function is the complete set of possible output values.

For example, consider the function f defined by $f(x) = x^2 - 4x + 5, \; x > 0$.

In completed-square form $x^2 - 4x + 5 = (x-2)^2 - 4 + 5$
$$= (x-2)^2 + 1$$

which is a minimum when $x = 2$.

The value $x = 2$ is in the domain of f so the minimum value of $f(x)$ is $f(2)$ which is 1.
Hence the range is the set of all the real numbers greater than or equal to 1.
We can write the range as $f(x) \geq 1$.

A sketch graph illustrates this.

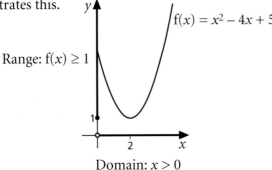

Range: $f(x) \geq 1$

$f(x) = x^2 - 4x + 5$

Domain: $x > 0$

Example 2

Draw a sketch graph of the function defined by $f(x) = 2x + 1, \; -3 < x < 4$.
State its range.

Solution

The graph is a straight line so find the end-points.

When $x = -3$, $2x + 1 = 2 \times -3 + 1 = -5$

When $x = 4$, $2x + 1 = 2 \times 4 + 1 = 9$

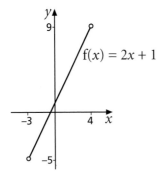

$f(x) = 2x + 1$

The end-points $(-3, -5)$ and $(4, 9)$ are not included in the graph and so are shown by open circles.

The range is $-5 < f(x) < 9$.

Example 3

Draw a sketch graph of the function defined by $g(t) = 3^t$, $t \in \mathbb{R}$.
State its range.

Solution

$g(0) = 3^0 = 1$

As t becomes large and positive, $g(t)$ gets larger very quickly.
As t becomes large and negative, $g(t)$ gets smaller and smaller, getting closer and closer to 0.

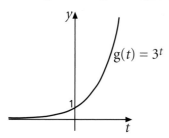

The graph gets closer and closer to the horizontal axis but does not reach it,
so the range is $g(t) > 0$.

Example 4

Draw a sketch graph of the function defined by $f(x) = \dfrac{6}{x+1}$, $x > 1$.
State its range.

Solution

When $x = 1$, $\dfrac{6}{x+1} = \dfrac{6}{2} = 3$ so $(1, 3)$ is the left-hand end point.
As x becomes large and positive, $f(x)$ gets smaller and smaller, getting closer and closer to 0.

A sketch is

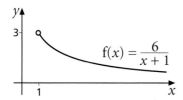

The graph gets closer and closer to the horizontal axis but does not reach it so
the range is $0 < f(x) < 3$.

Exercise C (answers p 162)

1 For each function below, draw a sketch graph and state the range of the function.

(a) $f(x) = x + 3$, $x \geq 1$

(b) $g(x) = x^2 + 3$, $x \in \mathbb{R}$

(c) $h(x) = 1 - x^2$, $x \in \mathbb{R}$

(d) $f(x) = \sin x$, $0 \leq x \leq 2\pi$

(e) $g(x) = 3^x$, $x > -1$

(f) $h(x) = \dfrac{4}{x-1}$, $x \geq 2$

(g) $f(x) = (x - 2)^2 + 3$, $x \in \mathbb{R}$

(h) $g(x) = (x + 1)^2 - 2$, $x \geq 0$

2 (a) Write the expression $x^2 - 6x + 10$ in completed-square form.

  (b) (i) Sketch the graph of $y = f(x)$ where f is defined by

  f: $x \mapsto x^2 - 6x + 10, \ x \le 2$

  (ii) Find the range of f.

3 (a) Sketch the graph of the function $g(\theta) = \cos(2\theta)$ where the domain of g is $0 \le \theta \le \frac{1}{2}\pi$.

  (b) State the range of g.

4 The function f is defined by $f(x) = x^2 + 2x + 6, \ x \in \mathbb{R}$.

  (a) Find the range of f.

  (b) Hence show that the equation $f(x) = 3$ has no real solution.

5 The function h is defined as $h(t) = t^2 - 4t - 5, \ -1 \le t \le 6$.

  (a) Show that this function is many–one.

  (b) Find the range of h.

6 The function h is defined by

  $$h(x) = \frac{1}{x} + 2, \ x \in \mathbb{R}, \ x \ne 0$$

  A sketch of its graph is shown.

  (a) Solve these equations.

  (i) $h(x) = 2\frac{1}{3}$ **(ii)** $h(x) = 3$

  (iii) $h(x) = 4$ **(iv)** $h(x) = 1$

  (b) Explain why the equation $h(x) = 2$ has no solution.

  (c) Which set of values below is the range of h?

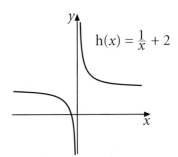

$h(x) = \frac{1}{x} + 2$

$h(x) \in \mathbb{R}$	$h(x) \in \mathbb{R}, \ h(x) \ne 2$	$h(x) \in \mathbb{R}, \ h(x) \ne 0$

7 What is the range of each function?

  (a) $f(x) = \frac{1}{x} - 5, \ x \in \mathbb{R}, \ x \ne 0$ **(b)** $f(x) = \frac{4}{x} + 3, \ x \in \mathbb{R}, \ x \ne 0$

8 The function g is defined by

  $$g(x) = \frac{1}{x}, \ x \in \mathbb{R}, \ x \ge 2$$

  Sketch the graph of the function and find its range.

9 The function f is defined by

  f: $t \mapsto 2^{-t}, \ t > -2$

  Find the range of f.

10 The function g is defined by
$$g(x) = \frac{1}{\sqrt{x}}, \quad x \geq \tfrac{1}{9}$$

(a) Solve the equation $g(x) = 2$.

(b) Find the range of g.

11 The function h is defined by
$$h(x) = 3 + \frac{4}{x-1}, \quad x < 1$$

Find the range of h.

D Composite functions (answers p 163)

D1 A gas meter indicates the amount of gas in cubic feet used by a consumer. The number of therms of heat from x cubic feet of gas is given by the function f where
$$f(x) = 1.034x, \quad x \geq 0$$

A particular gas company's charge in £ for t therms is given by the function g where
$$g(t) = 15 + 0.4t, \quad t \geq 0$$

(a) How many therms of heat are produced from 500 cubic feet of gas?

(b) What is the cost of using 100 therms?

(c) Find the cost of using these amounts of gas from this gas company.

    (i) 100 cubic feet   (ii) 400 cubic feet   (iii) 1000 cubic feet

To find the cost of using, say, 50 cubic feet of gas you must use both functions: first f and then g. This can be illustrated by the following diagram.

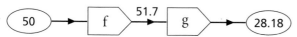

We can write this in stages as $gf(50) = g(f(50)) = g(51.7) = 28.18$.

D2 Evaluate $gf(200)$.

For an input of x cubic feet of gas we have the following diagram.

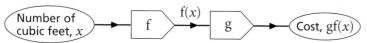

K   The function gf is called a **composite** function as it is a composition of two functions, f and g. gf means first f and then g.

D **D3** (a) Find the rule, in its simplest form, for the function gf where $gf(x)$ is the cost of using x cubic feet of gas.

    (b) Use your rule to find the cost of 550 cubic feet of gas.

Example 5

Functions f and g are defined for all real values of x by $\quad f(x) = x^2 + 1$
$$\text{and} \quad g(x) = 5x - 7$$

Calculate fg(2), gf(2) and ff(2).

Solution

$\text{fg}(2) = \text{f}(\text{g}(2)) = \text{f}(5 \times 2 - 7) = \text{f}(3) = 3^2 + 1 = 10$

$\text{gf}(2) = \text{g}(\text{f}(2)) = \text{g}(2^2 + 1) = \text{g}(5) = 5 \times 5 - 7 = 18$

$\text{ff}(2) = \text{f}(\text{f}(2)) = \text{f}(2^2 + 1) = \text{f}(5) = 5^2 + 1 = 26$

Example 6

Functions f and g are defined for all real values of x by $\quad f(x) = (x + 4)^2$
$$\text{and} \quad g(x) = 2x - 1$$

Find an expression for fg(x).

Solution

$\text{fg}(x) = \text{f}(\text{g}(x)) = \text{f}(2x - 1)$
$\qquad\qquad = ((2x - 1) + 4)^2$
$\qquad\qquad = (2x + 3)^2$

Example 7

Functions g and h are defined by $\quad g(x) = \dfrac{6}{x}, \ x \in \mathbb{R}, \ x \neq 0$

$$\text{and} \quad h(x) = \dfrac{2}{x + 5}, \ x \in \mathbb{R}, \ x \neq -5$$

Find and simplify an expression for gh(x).

Solution

$\text{gh}(x) = \text{g}(\text{h}(x)) = \text{g}\left(\dfrac{2}{x+5}\right) = \dfrac{6}{\left(\dfrac{2}{x+5}\right)}$

$\qquad\qquad\qquad = 6 \times \dfrac{x+5}{2}$

$\qquad\qquad\qquad = 3(x + 5)$

D4 Functions f and g are defined for all real values of x by
$$f(x) = 10 - x^2$$
$$\text{and} \quad g(x) = 3x + 2$$

(a) Evaluate these.

    (i) fg(0)   (ii) gf(4)   (iii) ff(−2)

(b) Find an expression for each of these.

    (i) gf(x)   (ii) fg(x)   (iii) gg(x)

Exercise D (answers p 163)

1 Functions f and g are defined for all real values of x by
$$f(x) = x^2$$
and $$g(x) = 3x + 1$$

(a) Evaluate these.

(i) fg(2) (ii) gf(2) (iii) gg(2)

(b) Find an expression for gf(x).

(c) Show that $fg(x) = 9x^2 + 6x + 1$.

(d) (i) Find an expression, in its simplest form, for gg(x).

(ii) Use your result to evaluate gg(−1).

(iii) Solve the equation gg(x) = 49.

2 For each pair of rules below, find expressions for fg(x) and gf(x).

(a) $f(x) = 2x + 3$ (b) $f(x) = x^2$ (c) $f(x) = 3x + 2$ (d) $f(x) = 1 - x^2$

$g(x) = x^3$ $g(x) = \dfrac{1}{x+1}$ $g(x) = 5 - x$ $g(x) = 1 - 2x$

3 Functions f and g are defined for all real values of x by
$$f: x \mapsto x^2 - 4x + 1$$
$$g: x \mapsto kx + 5, \text{ where } k \text{ is a constant}$$

Given that gf(1) = 2, find the value of k.

4 Functions f and g are defined by
$$f(x) = \frac{1}{x}, \ x > 0$$

$$g(x) = \frac{1}{3x - 1}, \ x > 1$$

Find an expression for fg(x) and write down the domain of fg.

5 Functions g and h are defined by
$$g: x \mapsto 2 - \frac{6}{x}, \ x > 0$$
$$h: x \mapsto x + 3, \ x > 0$$

(a) Show that $gh(x) = \dfrac{2x}{x + 3}$.

(b) Solve gh(x) = 1.

6 Functions f and g are defined by
$$f: x \mapsto x - 10, \ x \in \mathbb{R}$$
$$g: x \mapsto \sqrt{x}, \ x \geq 0$$

(a) Find f(1).

(b) Explain why the composite function gf cannot be formed.

7 Functions f and g are defined by

$$f(x) = \frac{1}{(x-1)(x+3)}, \quad x > 1$$

$$g(x) = \frac{5}{x}, \quad x > 0$$

(a) Solve $gf(x) = 25$.

(b) Write down the domain of gf.

8 A theatre manager notices that if he raises the temperature on the central heating thermostat he can increase the sales of ice cream in the interval.

He observes that the proportion of the audience buying ices is given by the function P where

$$P(c) = 1 - \frac{10}{c}, \quad 15 \le c \le 25$$

and where c is the temperature in degrees Celsius.

(a) What proportion of the audience buys ice cream when the temperature is $15\,°C$?

(b) At what temperature will half of the audience buy ices?

(c) What is the range of the function P?

The function $f(t) = \frac{5}{9}(t - 32)$ gives the temperature in degrees Celsius where t is the temperature in degrees Fahrenheit.

(d) What proportion of the audience buys ice cream when the temperature is $65\,°F$?

(e) Find the rule for the composite function that determines the proportion of the audience that will buy ices when the temperature is $t\,°F$.

(f) One evening, 55% of the audience buy ices.
What is the temperature in degrees Fahrenheit?

*(g) Work out an appropriate domain for the function Pf.

E Inverse functions (answers p 163)

E1 Functions f and g are defined for all real values of x by

$$f(x) = 2x + 5$$

$$g(x) = \frac{x-5}{2}$$

(a) Evaluate $gf(5)$ and $fg(5)$.

(b) Evaluate $gf(-3)$ and $fg(-3)$.

(c) Find an expression for $gf(x)$.

K A function that reverses, or 'undoes', the effect of f is its **inverse** and is denoted by f^{-1}.
So $f^{-1}f(x) = x$. It is also true that $ff^{-1}(x) = x$.

(In this context, f^{-1} does not mean $\frac{1}{f}$.)

So we can write $f(x) = 2x + 5$ and $f^{-1}(x) = \frac{x-5}{2}$.

E2 The rule for the function f is $f(x) = 3x - 1$.

    (a) (i) Show that $f(7) = 20$.

       (ii) Hence write down the value of $f^{-1}(20)$.

    (b) Which of the rules below is the rule for the inverse of f?

A $f^{-1}(x) = 3x + 1$ **B** $f^{-1}(x) = \frac{1}{3}x + 1$

C $f^{-1}(x) = \dfrac{x+1}{3}$ **D** $f^{-1}(x) = \dfrac{x-1}{3}$

One way to find a rule for the inverse of a function is to write the function in terms of x and y and then rearrange to obtain a rule for x in terms of y.

For example, to find the rule for the inverse of $g(x) = 2x - 1$
first write the rule as $y = 2x - 1$
then rearrange to obtain $y + 1 = 2x$

$$\Rightarrow \tfrac{1}{2}(y + 1) = x$$

So we can write the inverse as $g^{-1}(y) = \tfrac{1}{2}(y + 1)$

We could write the inverse rule as, say, $g^{-1}(k) = \tfrac{1}{2}(k + 1)$ or $g^{-1}(p) = \tfrac{1}{2}(p + 1)$ or use any other letter we choose.

We usually use x which gives $g^{-1}(x) = \tfrac{1}{2}(x + 1)$ for the inverse rule.

E3 For each rule, find an expression for $f^{-1}(x)$, where f^{-1} is the inverse of f.

    (a) $f(x) = 4x + 3$ **(b)** $f(x) = \tfrac{1}{5}x + 3$ **(c)** $f: x \mapsto 2(x - 7)$

E4 For each rule, find an expression for $g^{-1}(x)$, where g^{-1} is the inverse of g.

    (a) $g(x) = 10 - x$ **(b)** $g: x \mapsto -x$ **(c)** $g(x) = \dfrac{1}{x}$

An inverse exists for any one–one function.
If a function is many–one, then an inverse function does not exist.

For example, for the many–one function $f(x) = x^2$ we have
two different inputs (-2 and 2) that generate the same output (4)
so reversing the effect of f gives two possible outputs for
an input of 4.

A function must give exactly one output for each input
so an inverse function for $f(x) = x^2$ does not exist.

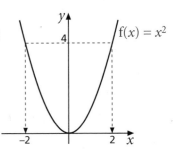

E5 Which of these functions does not have an inverse?

    A $f(x) = x^2 - 5,\ -3 \le x \le 3$ **B** $f(x) = x^2 - 5,\ 1 \le x \le 6$

E6 The domain of a function g is $-2 \le x \le 6$. The graph of $y = g(x)$ is shown.

(a) Write down the value of

(i) $g(3)$ (ii) $g^{-1}(3)$ (iii) $g^{-1}(4)$ (iv) $g^{-1}(-1)$

(b) Why is it not possible to evaluate $g^{-1}(10)$?

(c) What is the domain of the function g^{-1}?

(d) Write down the range of the function g^{-1}.

(e) Sketch the graph of the function $y = g^{-1}(x)$.

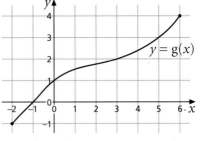

 The domain of an inverse function f^{-1} is the range of the function f.

E7 For each function below,

(i) using the same scale on the x- and y-axes, sketch the graph of $y = f(x)$

(ii) find an expression for $f^{-1}(x)$

(iii) find the domain of f^{-1}

(iv) add the graph of $y = f^{-1}(x)$ to your sketch of $y = f(x)$

(a) $f(x) = 2x + 1, \ x \ge -2$ (b) $f(x) = \frac{1}{4}(x - 2), \ x < 10$

(c) $f(x) = x^2, \ x > 1$ (d) $f: x \mapsto x^3 + 1, \ 0 \le x \le 2$

 Using the same scale on the x- and y-axes, the graphs of a function and its inverse have reflection symmetry in the line $y = x$.

Example 8

A function f is defined for $x \ge 0$ by $f(x) = (x + 1)^2 + 2$.
Find an expression for the inverse $f^{-1}(x)$.
Sketch the graph of the inverse function f^{-1} and state its domain.

Solution

First write the function in terms of x and y.
$$y = (x + 1)^2 + 2$$
$$\Rightarrow \quad y - 2 = (x + 1)^2$$

Take the positive square root as $x \ge 0$.
$$\sqrt{y - 2} = x + 1$$
$$\Rightarrow \quad \sqrt{y - 2} - 1 = x$$

We need an expression for $f^{-1}(x)$.
So $f^{-1}(x) = \sqrt{x - 2} - 1$

You can first sketch the graph of $y = (x + 1)^2 + 2$ for $x \in \mathbb{R}$, remembering to use the same scale on the x- and y-axes.

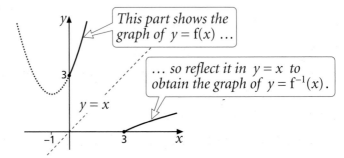

This part shows the graph of $y = f(x)$ ...

... so reflect it in $y = x$ to obtain the graph of $y = f^{-1}(x)$.

$y = x$

The graph is

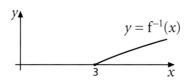

The range of f is $f(x) \geq 3$ so the domain of f^{-1} is $x \geq 3$.

Example 9

A function g has the rule defined by $g: x \mapsto \dfrac{3}{2x-1}$, $x \geq 1$

Find an expression for the inverse $g^{-1}(x)$ and find the domain of g^{-1}.

Solution

First write the rule in terms of x and y. $y = \dfrac{3}{2x-1}$

Multiply both sides by $(2x-1)$. $y(2x-1) = 3$

Expand the brackets. $2xy - y = 3$

$\Rightarrow \qquad 2xy = 3 + y$

$\Rightarrow \qquad x = \dfrac{3+y}{2y}$

We need an expression for $g^{-1}(x)$.   So $g^{-1}(x) = \dfrac{3+x}{2x}$

The domain of g^{-1} is the range of g.

When $x = 1$, $g(x) = \dfrac{3}{2x-1} = 3$ and when $x > 1$, $\dfrac{3}{2x-1} < 3$.

As x becomes large and positive, $\dfrac{3}{2x-1}$ gets smaller and smaller, getting closer and closer to 0.

So the range of g is $0 < g(x) \leq 3$.

Hence, the domain of g^{-1} is $0 < x \leq 3$.

Exercise E (answers p 164)

1 Function f is defined by $f(x) = 5x + 1$, $x \in \mathbb{R}$.
Find an expression for $f^{-1}(x)$.

2 Function g is defined by $g(x) = \frac{1}{4}x - 3$, $0 \leq x \leq 16$.
(a) Find an expression for $g^{-1}(x)$.
(b) Find the domain and range of g^{-1}.

3 The function $f(t) = \frac{5}{9}(t - 32)$ gives the temperature in degrees Celsius where t is the temperature in degrees Fahrenheit. The domain of the function is $t \geq -459.4$.
(a) Find an expression for $f^{-1}(t)$.
(b) What is the domain of f^{-1}?
(c) Convert $-70\,^{\circ}$C into a temperature in degrees Fahrenheit.

4 Explain why no inverse exists for function f where f is defined by $f(x) = x^2$, $x \geq -4$.

5 Function h is defined by h: $x \mapsto 2 - 3x$, $x > 0$.

   (a) Find an expression for $h^{-1}(x)$.

   (b) What is the domain of h^{-1}?

   (c) Solve the equation $h^{-1}(x) = h(x)$.

6 Function f is defined by $f(x) = (x - 2)^2 - 5$, $x > 2$.

   (a) Find an expression for $f^{-1}(x)$.

   (b) What is the domain of f^{-1}?

7 Function g is defined by g: $x \mapsto x^2 + 6x + 10$, $x > -2$.

   (a) Write $x^2 + 6x + 10$ in completed-square form.

   (b) Find an expression for $g^{-1}(x)$.

8 Find an expression for each inverse function $f^{-1}(x)$ and write down its domain.

   (a) $f(x) = \sqrt{x} + 2$, $x \geq 0$ **(b)** $f(x) = x^3 - 5$, $x \in \mathbb{R}$

   (c) $f(x) = \dfrac{2}{3 + x}$, $x > 0$ **(d)** $f(x) = \dfrac{5}{x} - 4$, $x < 0$

9 The graph sketched is of the function defined by
$$f: x \mapsto 3^{-x} - \tfrac{1}{3}, \ x \geq 0$$

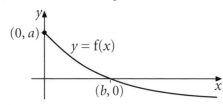

   (a) Work out the values of a and b.

   (b) Sketch the curve with equation $y = f^{-1}(x)$.

   (c) What is the domain of f^{-1}?

   (d) What is the range of f^{-1}?

10 Function f is defined by $f(x) = \dfrac{2x + 3}{x - 2}$.

   Show that $f^{-1}(x) = f(x)$ for $x \in \mathbb{R}$, $x \neq 2$.

11 Function g is defined as g: $x \mapsto x^2 - 2x + 5$, $x > 1$.

   (a) Find an expression for $g^{-1}(x)$.

   (b) Sketch the graphs of $y = g(x)$ and $y = g^{-1}(x)$.

   (c) State the domain and range of g^{-1}.

   (d) Solve the equation $g(x) = 7$, leaving your solution in surd form.

   (e) Show that the equation $g^{-1}(x) = g(x)$ has no real solutions.

Key points

- A **function** is a rule or process that generates exactly one output for every input. A definition of a function consists of

    a **rule** that tells you how values of the function are calculated

    a **domain** that is the set of values to which the rule is applied (p 21)

- An alternative notation for the rule of a function is to use an arrow. For example, rather than writing $f(x) = x^2$ we can write $f: x \mapsto x^2$. (p 23)

- A **many–one function** is a function where two or more different inputs generate the same output. Any function where each output can be generated by only one input is a **one–one function**. (p 24)

- The **range** of a function is the complete set of possible output values. (p 26)

- The function gf is called a **composite** function and tells you to 'do f first and then g': $gf(x) = g(f(x))$. (p 29)

- A function that reverses, or 'undoes', the effect of f is its **inverse** and is denoted by f^{-1}. So $f^{-1}f(x) = x$.
  Only one–one functions have inverses.
  The domain of an inverse function f^{-1} is the range of the function f. (pp 32–34)

- One way to find a rule for the inverse of a function is to write the function in terms of x and y and then rearrange to obtain a rule for x in terms of y. (p 33)

- Using the same scale on the x- and y-axes, the graphs of a function and its inverse have reflection symmetry in the line $y = x$. (p 34)

Mixed questions (answers p 165)

1 The function f is defined for all real values of x by $f(x) = (x + 2)(x - 2)$.

   (a) Sketch the graph of $y = f(x)$, showing where the graph crosses both axes.

   (b) Find the range of f.

   (c) Explain why the function f does not have an inverse.

2 The function f has domain $x \geq 0$ and is defined by $f(x) = (x + 1)^2 - 3$.

   (a) Find the range of f.

   (b) Explain why the equation $f(x) = -3$ has no solution.

   (c) (i) Write down the domain of f^{-1} where f^{-1} is the inverse of f.

   (ii) Find an expression for $f^{-1}(x)$.

3 The function f is defined by

   $f: t \mapsto 5^t, \ t \leq 2$

   Find the range of f.

4 $f(x) = \dfrac{2}{x-1} - \dfrac{6}{(x-1)(2x+1)}, \ x > 1.$

   (a) Prove that $f(x) = \dfrac{4}{2x+1}.$

   (b) Find the range of f.

   (c) Find $f^{-1}(x)$.

   (d) Find the range of f^{-1}.   Edexcel

5 Functions f and g are defined by

$$f: x \mapsto \dfrac{x}{x-3}, \ x \in \mathbb{R}, \ x \neq 3$$

$$g: x \mapsto \dfrac{1}{2x-1}, \ x \in \mathbb{R}, \ x \neq \tfrac{1}{2}$$

   (a) **(i)** Show that $gf(x) = 1 - \dfrac{6}{x+3}.$

        (ii) Solve $gf(x) = 7.$

   (b) **(i)** Find an expression for $f^{-1}(x).$

        (ii) Find the domain of $f^{-1}.$

6 Function k is defined by $k(x) = \sqrt{x-1} + 3, \ x \geq 1.$

   (a) Find an expression for $k^{-1}(x).$

   (b) Explain why the x-coordinate of any point of intersection of the graphs of $y = k(x)$ and $y = k^{-1}(x)$ must satisfy the equation $k^{-1}(x) = x.$
Hence, find the x-coordinate of any point of intersection.

Test yourself (answers p 166)

1 Functions f and g are defined for all real values of x by

$$f: x \mapsto 1 + 3x$$

$$g: x \mapsto (x-1)^3$$

   (a) The composite function gf is defined for all real values of x.
Find $gf(x)$, expressing your answer in its simplest form.

   (b) The sketch shows the graph of $y = g(x)$.

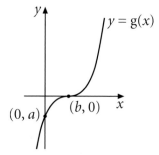

        (i) What are the values of a and b?

        (ii) Copy the graph of $y = g(x)$ and on the same axes sketch the graph of $y = g^{-1}(x)$.

   (c) Find an expression for $g^{-1}(x)$.

2 One of the following sketch graphs does not represent a function.
State which one this is and give a reason for your answer.

A

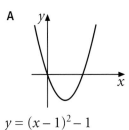

$y = (x-1)^2 - 1$

B

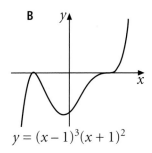

$y = (x-1)^3(x+1)^2$

C

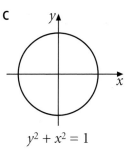

$y^2 + x^2 = 1$

3 Function f is defined by $f(x) = 2x^2, \ x \geq 0$.
(a) What is the domain of $f^{-1}(x)$?
(b) The graph of $y = f(x)$ and the graph of $y = f^{-1}(x)$ intersect at two points.
Find the x-coordinates of these two points.

4 The function h is defined for all real numbers by $h(x) = 8 - 3x^2$.
(a) Find the range of h.
(b) Solve the equation $h(x) = 5$.
(c) Explain why the function h does not have an inverse.

5 Functions f and g are defined by
$$f: x \mapsto \frac{10}{3+x}, \ x > 0$$
$$g: x \mapsto \frac{5}{x}, \ x \in \mathbb{R}, \ x \neq 0$$
(a) (i) Find gf(x), giving your answer in its simplest form.
(ii) What is the domain of gf?
(b) What is the range of f?
(c) $f^{-1}(x)$ can be written in the form $\frac{A}{x} + B$ where A and B are constants.
(i) Find the values of A and B.
(ii) Solve the equation $f^{-1}(x) = f(x)$.

6 Functions f and g are defined for all real values of x by
$$f(x) = x^2 - 4x + 7$$
$$g(x) = x + k, \ \text{where } k \text{ is a positive constant}$$
(a) Find the range of f.
(b) Given that $fg(3) = 12$, find the value of k.
(c) Solve the equation $gf(x) = 14$.

3 The modulus function

In this chapter you will learn
- what is meant by 'the modulus function'
- how to draw the graph of a modulus function
- how to solve an equation or inequality that involves a modulus function and relate the solution to the graphs

A Introducing the modulus function

$|x|$ is the symbol for 'the modulus of x' or 'the absolute value of x'.
The modulus or absolute value of a real number can be thought of as its 'distance' from 0 and is always positive.

For example, $|4| = 4$ and $|-2| = 2$.

Many graphic calculators use 'Abs (x)' for $|x|$.

Example 1

Given that $f(x) = |2x - 5|$, find the value of $f(1)$.

Solution

$f(1) = |2 \times 1 - 5| = |-3| = 3$

Exercise A (answers p 166)

1 Evaluate these.

    (a) $|-3| + 1$ **(b)** $|-3 + 1|$ **(c)** $|7 - 1|$ **(d)** $|1 - 7|$

2 A function is defined by $f(x) = |2x - 3|$.
    Evaluate these.

    (a) $f(1)$ **(b)** $f(2)$ **(c)** $f(0)$ **(d)** $f(6)$ **(e)** $f(-3)$

3 A function is defined by $g(x) = |4 - x| + 1$.
    Evaluate these.

    (a) $g(3)$ **(b)** $g(6)$ **(c)** $g(-1)$ **(d)** $g(10)$ **(e)** $g\left(5\tfrac{1}{2}\right)$

4 A function is defined by $h(x) = |x^2 - 5|$.
    Evaluate these.

    (a) $h(1)$ **(b)** $h(3)$ **(c)** $h(0)$ **(d)** $h(-2)$ **(e)** $h(-4)$

5 Functions f and g are defined by $f(x) = |x - 3|$ and $g(x) = x + 1$.
    Evaluate these.

    (a) $fg(0)$ **(b)** $gf(0)$ **(c)** $ff(1)$ **(d)** $gf(-2)$ **(e)** $fg(2)$

B Graphs (answers p 166)

You can check each graph on a graph plotter.

B1 (a) Copy and complete this table of values.

x	−3	−2	−1	0	1	2	3
$\|x\|$							

(b) Evaluate these.

(i) $\left|\frac{1}{2}\right|$ **(ii)** $\left|-\frac{3}{4}\right|$ **(iii)** $\left|-2.25\right|$

(c) Draw the graph of $y = |x|$.

B2 (a) Copy and complete this table of values.

x	−3	−2	−1	0	1	2	3
$\|x + 1\|$							

(b) Draw the graph of $y = |x + 1|$.

B3 (a) Draw up a table of values for the equation $y = |x^2 - 4|$.

(b) Draw the graph of $y = |x^2 - 4|$.

B4 Match each equation with its sketch graph.

(a) $y = |x - 2|$ **(b)** $y = |x + 5|$ **(c)** $y = |2x|$

(d) $y = \left|\frac{1}{x}\right|$ **(e)** $y = |x^2 - 3|$ **(f)** $y = |x^3 - x|$

A
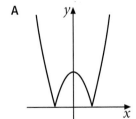

B
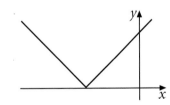

C
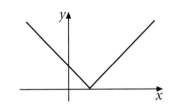

D
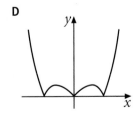

E
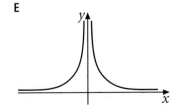

F
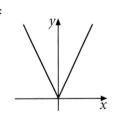

B5 Show that the graph of $y = |x - 2|$ is the same as the graph of $y = |2 - x|$.

One method to sketch the graph of $y = |2x - 1|$ is shown below.

First sketch the graph of $y = 2x - 1$.

Identify which parts of the graph lie beneath the x-axis (showing where the values of $2x - 1$ are negative).

Change the negative values to positive ones by reflecting in the x-axis.

Sketch the final graph of $y = |2x - 1|$.

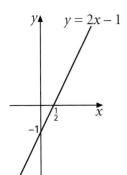
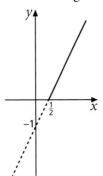
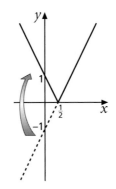
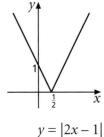

This method can be used to sketch the graph of the modulus of any expression.

Example 2

A function is defined by $f(x) = x^2 - 4x + 3$.
Draw the graph of $y = |f(x)|$ showing clearly where the graph meets each axis.

Solution

$f(0) = 3$ so the graph of $y = f(x)$ cuts the y-axis at $(0, 3)$.

$x^2 - 4x + 3 = (x - 3)(x - 1)$
So the graph of $y = f(x)$ cuts the x-axis at $(1, 0)$ and $(3, 0)$.

As there is a vertical line of symmetry, the vertex is halfway between 1 and 3 at $x = 2$.
$f(2) = -1$ so the vertex of the graph $y = f(x)$ is $(2, -1)$.

The graph of $y = f(x)$ is

So the graph of $y = |f(x)|$ is

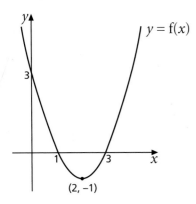
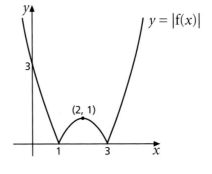

Exercise B (answers p 167)

In these questions, show clearly the points where each graph meets the axes.
Where possible, check your graph on a graph plotter.

1 Sketch the graph of each equation, using a separate diagram for each one.

(a) $y = |x + 5|$ (b) $y = |3x - 5|$ (c) $y = |2x + 8|$

2 Sketch the graph of each equation, using a separate diagram for each one.

(a) $y = |x^2 - 1|$ (b) $y = |x^2 - x|$ (c) $y = |x^2 - x - 6|$

3 Show that the graph of $y = |x^2 - 9|$ is the same as the graph of $y = |9 - x^2|$.

4 The diagram below shows a sketch of the curve with equation $y = f(x)$, $-2 \leq x \leq 4$.

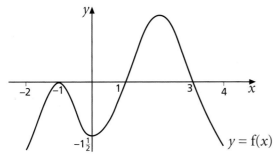

Draw a sketch of the curve with equation $y = |f(x)|$.

5 Sketch the graph of $y = |\sin x|$, $-2\pi \leq x \leq 2\pi$.

6 A function is defined as $g(x) = (x + 1)(x - 2)(x - 3)$, $-1 \leq x \leq 3$.
Sketch the graph of $y = |g(x)|$.

7 A function is defined as $h(x) = 2^x - 1$, $-2 \leq x \leq 3$.
Sketch the graph of $y = |h(x)|$.

8 The function f is defined as f: $x \mapsto |x + k|$, $x \in \mathbb{R}$, where k is a positive constant.
Sketch the graph of $y = f(x)$.

9 Each of the graphs below has a vertical line of symmetry.
Determine the equation of each graph.

(a)

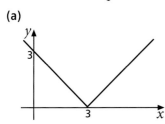

(b)

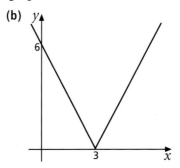

(c)

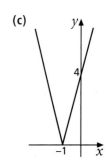

C Equations and inequalities (answers p 167)

C1 Show that $x = -4$ is a solution of the equation $|x + 1| = 3$.
Find another solution of this equation.

C2 Solve these equations.

(a) $|x + 4| = 5$ (b) $|x - 1| = 5$ (c) $|x + 3| = 1$ (d) $|x + 6| = 0$

C3 Explain how you know that the equation $|x + 2| = -1$ has no solution.

An equation such as $|2x + 1| = 3$ is straightforward to solve.
There are two possibilities: either $2x + 1 = 3$ or $2x + 1 = -3$.

$$2x + 1 = 3 \quad \Rightarrow \quad x = 1$$
$$\text{and } 2x + 1 = -3 \quad \Rightarrow \quad x = -2$$

So there are two solutions, $x = 1$ and $x = -2$.

A graph illustrates this.

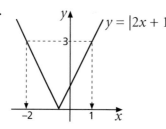

C4 Solve these equations.

(a) $|2x + 3| = 5$ (b) $|3x + 1| = 2$ (c) $|2x - 5| = 4$

D   **C5** Can the equation $|3x + 2| = x$ be solved?

We can see that the equation $|2x + 1| = x$ does not have a solution when we
sketch the graphs of $y = |2x + 1|$ and $y = x$ on the same axes.

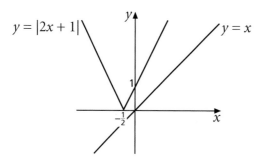

The graphs do not intersect so the equation $|2x + 1| = x$ cannot have a solution.

C6 (a) By sketching graphs, show that the equation $|2x + 1| = -x$ has two solutions.

    (b) Find these solutions.

When solving an equation that involves a modulus function, it often helps to start with a sketch.

For example, we can see from a sketch that the equation $|2x - 1| = x$ has two solutions (as the graphs of $y = |2x - 1|$ and $y = x$ intersect at two points).

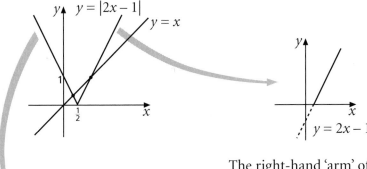

The right-hand 'arm' of the graph of $y = |2x - 1|$ is part of the graph of $y = 2x - 1$.

So for the point of intersection on this arm we have $2x - 1 = x$, which has a solution $x = 1$.

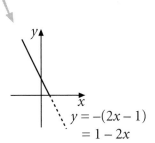

The left-hand arm of the graph of $y = |2x - 1|$ is part of the graph of $y = -(2x - 1)$, which is $y = 1 - 2x$.

So for the point of intersection on this arm we have $1 - 2x = x$, which has a solution $x = \frac{1}{3}$.

So the solutions of the equation $|2x - 1| = x$ are $x = 1$ and $x = \frac{1}{3}$.

C7 Solve the equation $|2x - 5| = x + 1$.

C8 (a) Draw a sketch of $y = |x^2 - 9|$.
   (b) Use your sketch to write down the number of solutions of each equation below.
      (i) $|x^2 - 9| = 7$ (ii) $|x^2 - 9| = -3$ (iii) $|x^2 - 9| = 16$
   (c) Find all the solutions of the equation $|x^2 - 9| = 7$.

C9 (a) On the same diagram, draw sketches of $y = |3x - 6|$ and $y = |x|$.
   (b) Find all the solutions of the equation $|3x - 6| = |x|$.
   (c) Solve the inequality $|3x - 6| < |x|$.

C10 Show that the equation $|x^2 - 1| = x - 2$ has no solutions.

C11 Solve the inequality $|4x + 1| \geq 3$.

C12 Solve the equation $|x| = 12 - x^2$.

Example 3

A function is defined by $f(x) = (x + 1)(x - 5)$.
Solve $|f(x)| = 8$.

Solution

A sketch shows the equation will have four solutions.

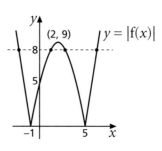

For $|f(x)| = 8$, either $f(x) = 8$ or $f(x) = -8$, so solve
$(x + 1)(x - 5) = 8$ and $(x + 1)(x - 5) = -8$.

$$(x + 1)(x - 5) = 8$$
$$x^2 - 4x - 5 = 8$$
$$\Rightarrow \quad x^2 - 4x - 13 = 0$$
$$\Rightarrow \quad (x - 2)^2 - 17 = 0$$
$$\Rightarrow \quad (x - 2)^2 = 17$$
$$\Rightarrow \quad x - 2 = \pm\sqrt{17}$$
$$\Rightarrow \quad x = 2 \pm \sqrt{17}$$

$$(x + 1)(x - 5) = -8$$
$$\Rightarrow \quad x^2 - 4x - 5 = -8$$
$$\Rightarrow \quad x^2 - 4x + 3 = 0$$
$$\Rightarrow \quad (x - 1)(x - 3) = 0$$
$$\Rightarrow \quad x = 1, 3$$

So the four solutions are $x = 2 + \sqrt{17},\, 2 - \sqrt{17},\, 1,\, 3$.

Example 4

Solve the equation $|3x - 5| = |x + 1|$.

Solution

A sketch shows that the equation has two solutions.

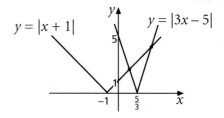

The right-hand solution is the intersection of $y = 3x - 5$ *and* $y = x + 1$, *so solve* $3x - 5 = x + 1$.

$$3x - 5 = x + 1 \quad \Rightarrow \quad 2x = 6$$
$$\Rightarrow \quad x = 3$$

The left-hand solution is the intersection of $y = -(3x - 5)$ *and* $y = x + 1$, *so solve* $-(3x - 5) = x + 1$.

$$-(3x - 5) = x + 1 \quad \Rightarrow \quad -3x + 5 = x + 1$$
$$\Rightarrow \quad 4x = 4$$
$$\Rightarrow \quad x = 1$$

So the two solutions are $x = 1, 3$.

Example 9

Solve the equation $x|x-1| = 1$.

Solution

$$x|x-1| = 1$$
$$\Rightarrow \quad |x-1| = \frac{1}{x}$$

A sketch of $y = |x-1|$ and $y = \dfrac{1}{x}$ shows that the equation

$|x-1| = \dfrac{1}{x}$ has one solution, which is when $x-1 = \dfrac{1}{x}$.

$$x-1 = \frac{1}{x}$$
$$\Rightarrow \quad x(x-1) = 1$$
$$\Rightarrow \quad x^2 - x - 1 = 0$$
$$\Rightarrow \quad x = \frac{-(-1) \pm \sqrt{(-1)^2 - 4 \times 1 \times (-1)}}{2}$$
$$\Rightarrow \quad x = \frac{1 \pm \sqrt{5}}{2}$$

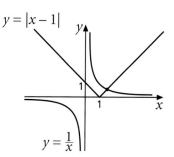

From the graph we can see that x is positive so the solution is $x = \frac{1}{2}(1 + \sqrt{5})$.

Exercise D (answers p 169)

1 Sketch the graph of $y = f(x)$ for each function.
State the range of f each time.

(a) $f(x) = |x| + 2$ (b) $f(x) = 3|x+1|$ (c) $f(x) = 2|x| + 3$

(d) $f(x) = |2x-1| - 3$ (e) $f(x) = 5 - |x|$ (f) $f(x) = 6 - |3x+1|$

2 Solve each equation.

(a) $2|x| = \frac{1}{2}x + 15$ (b) $|3x-1| - 5 = x$ (c) $4 - |x| = x$

3 The function f is defined for all real values of x by $f(x) = 8 - |2x+5|$.

(a) Solve the equation $f(x) = 1$.

(b) Show that there are no solutions to $f(x) = 9$.

(c) Solve the inequality $f(x) \le 2$.

4 The function g is defined for all real values of x by $g(x) = 1 + |3x-7|$.

(a) Solve the equation $g(x) = x$.

(b) Show that the equation $g(x) = x - 2$ has no solutions.

5 Solve the inequality $|x+2| < 3|x|$.

6 The function h is defined by $h(x) = |x-1| + 4$.
Solve $h(x) > 2x$.

7 (a) On the same set of axes, sketch the graphs of $y = |3x + 2|$ and $y = \dfrac{1}{x}$.

(b) Explain how your graphs show that there is only one solution of the equation $x|3x + 2| = 1$. Solve the equation.

8 The function f is defined by $f(x) = |x|,\ x \in \mathbb{R}$.
Show that f^{-1} does not exist.

9 Functions f and g are defined for all real values of x by
$$f(x) = |x^2 - 3|\ \text{and}\ g(x) = x + 2$$

(a) Write down, and simplify if necessary, expressions for $fg(x)$ and $gf(x)$.

(b) Sketch the graph of $y = gf(x)$.

10 Functions f and g are defined for all real values of x by
$$f(x) = |x + 1|\ \text{and}\ g(x) = x - 3$$

(a) Solve the equation $gf(x) = 5$.

(b) Solve the inequality $fg(x) < x$.

11 The function g is defined by $g(x) = |x| + x$.

(a) Sketch the graph of $y = g(x)$.

(b) Solve the equation $g(x) = 10$.

12 Solve the equation $|x| - 1 = |x^2 - 3|$.

E Graphing $y = f(|x|)$ (answers p 171)

E1 The function f is defined for all real values of x by $f(x) = x^2 - 3x$.

(a) Find the value of $f(4)$.

Let $y = f(|x|)$.

(b) Find the value of y when $x = -4$.

(c) Copy and complete this table.

x	-4	-3	-2	-1	0	1	2	3	4
$|x|$		3							
$y = f(|x|)$		0							

(d) (i) What can you say about the values of y when $x = k$ and $x = -k$?

(ii) What does this tell you about the graph of $y = f(|x|)$?

(iii) Sketch the graph of $y = f(|x|)$.

> **K** The graph of $y = f(|x|)$ has the y-axis as a line of symmetry.
>
> To sketch the graph of $y = f(|x|)$, first sketch the graph of $y = f(x)$ for $x \geq 0$.
> Now reflect this in the y-axis to obtain the graph for $x \leq 0$.
> Both halves together form the graph of $y = f(|x|)$.

Example 10

A function f is defined by $f(x) = \frac{1}{3}x^3 - x^2 - \frac{1}{3}x + 1$.
A sketch of $y = f(x)$ is shown.

Sketch the graph of $y = f(|x|)$.

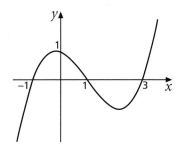

Solution

Sketch the graph of $y = f(x)$ for $x \geq 0$. ➡ *Reflect it in the y-axis to obtain the whole graph.*

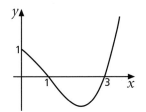 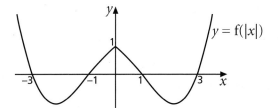

Example 11

Sketch the graph of $y = |x|^2 - 2|x|$.

Solution

Define f as $f(x) = x^2 - 2x$ so that $f(|x|) = |x|^2 - 2|x|$.
Hence $y = f(|x|)$ is the required graph.

Sketch the graph of $y = f(x)$. ➡ *Restrict the graph to $x \geq 0$.* ➡ *Reflect it in the y-axis.*

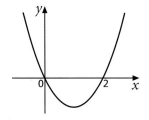 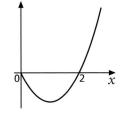 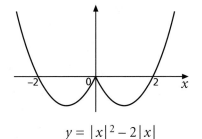

$x^2 - 2x = x(2 - x)$ *so*
$y = x^2 - 2x$ *is a parabola*
with 0 and 2 as x-intercepts.

$$y = |x|^2 - 2|x|$$

Exercise E (answers p 171)

1 A function f is defined by $f(x) = x^2 - x$.

    (a) Sketch graphs of   **(i)** $y = f(|x|)$ **(ii)** $y = |f(x)|$

    (b) For what range of values of x does $f(|x|) = |f(x)|$?

2 A sketch of $y = f(x)$ is shown.
Sketch the graph of $y = f(|x|)$.

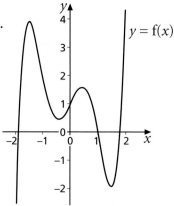

3 A function f is defined by $f(x) = (x - 2)(x - 4)$.

(a) Sketch the graph of $y = f(|x|)$.

(b) Solve these equations.

(i) $f(|x|) = 3$ (ii) $f(|x|) = 8$ (iii) $f(|x|) = 24$

4 Sketch the graph of each of these, indicating clearly the coordinates of any points where the graph meets the x- and y-axes.

(a) $y = (|x| - 1)^2$ (b) $y = 3^{|x|}$

(c) $y = |x|^2 - 4|x| - 5$ (d) $y = \sin|\theta|$, $-2\pi \leq \theta \leq 2\pi$

Key points

- $|x|$ is the symbol for 'the modulus of x' or 'the absolute value of x'. The modulus of a real number can be thought of as its 'distance' from 0 or its 'size' and is always positive. For example, $|4| = 4$ and $|-2| = 2$.   (p 40)

- The graph of $y = |x|$ is

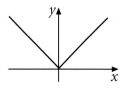

   (p 41)

- One method of sketching $y = |f(x)|$ is first to sketch the graph of $y = f(x)$. Take any part of the graph that is below the x-axis and reflect it in the x-axis to obtain the graph of $y = |f(x)|$.   (p 42)

- When solving an equation that involves a modulus function, it is usually a good idea to sketch graphs first to determine the number of solutions.   (p 45)

- One method of sketching $y = f(|x|)$ is to first sketch the graph of $y = f(x)$ for $x \geq 0$. Now reflect this in the y-axis to sketch the graph for $x \leq 0$. Both halves together form the graph of $y = f(|x|)$.   (pp 52–53)

1 The function f is defined for all real values of x by
$$f(x) = |x + 1| + 6$$

   (a) Sketch the graph of $y = f(x)$ and state the range of f.

   (b) Evaluate ff(-5).

   (c) Solve the equation $f(x) = 10$.

2 The function f is defined for all real values of x by
$$f(x) = |4x - 1| - 3$$

   (a) Sketch the graph of $y = f(x)$, showing clearly where the graph crosses the x- and y-axes.

   (b) Write down the range of f.

   (c) Solve the equation $f(x) = 2x$.

3 The function g is defined for all real values of x by
$$g(x) = 5 - |2x - 3|$$

   (a) (i) Sketch the graph of $y = g(x)$, showing clearly where the graph crosses the x- and y-axes.

   (ii) State the range of g.

   (b) (i) For which values of k does the equation $g(x) = k$ have no real roots?

   (ii) Solve the equation $g(x) = 2$.

   (c) Solve the inequality $g(x) > x$.

4 Solve the equation $|3x + 1| = |3x - 5|$.

5 (a) Sketch, on the same diagram, the graphs of
$$y = |2x + 1| \text{ and } y = 4 - x^2$$
   indicating the coordinates of any points where the graphs meet the axes.

   (b) Solve the equation $|2x + 1| = 4 - x^2$, giving the exact value of each root.

6 Solve the inequality $|x - 2| > 2|x + 1|$.

7 (a) Sketch the graph of $y = |2x + a|$, $a > 0$, showing the coordinates of the points where the graph meets the coordinate axes.

   (b) On the same axes, sketch the graph of $y = \dfrac{1}{x}$.

   (c) Explain how your graphs show that there is only one solution of the equation
$$x|2x + a| - 1 = 0$$

   (d) Find, using algebra, the value of x for which $x|2x + 1| - 1 = 0$.   Edexcel

8 The function f is defined for all non-zero real values of x by $f(x) = \dfrac{1}{x}$.
   Solve the equation $f(|x|) = 2$.

4 Transforming graphs

In this chapter you will
- revise how to transform the graph of $y = f(x)$ to obtain graphs of the form $y = af(x)$, $y = f(ax)$, $y = f(x) + a$ and $f(x + a)$
- learn how to combine transformations to obtain a new graph and a new equation

A Single transformations: revision (answers p 173)

If one graph is a transformation of another, then the equations of the two graphs are related.

For example, transforming the graph
of $y = x^2$ by translating it by $\begin{bmatrix} 0 \\ 3 \end{bmatrix}$ gives
a new graph with equation $y = x^2 + 3$.

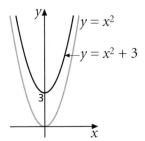

Function notation helps us think about how the equations of transformed graphs are related. For example, suppose the graph of $y = f(x)$ is transformed by a stretch of scale factor 2 in the x-direction. Thus the x-coordinate of each point is doubled.

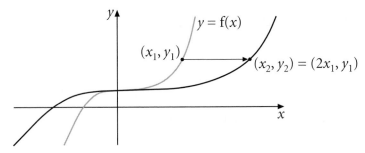

Let (x_1, y_1) be a point on $y = f(x)$ and let (x_2, y_2) be its image on the stretched curve.
Then $(x_2, y_2) = (2x_1, y_1)$, giving

$$x_2 = 2x_1 \text{ which rearranges to } x_1 = \tfrac{1}{2}x_2$$
$$\text{and } y_2 = y_1 \text{ which rearranges to } y_1 = y_2$$

We know that $y_1 = f(x_1)$, so it must be true that
$$y_2 = f\left(\tfrac{1}{2}x_2\right)$$
and so $y = f\left(\tfrac{1}{2}x\right)$ is the equation of the image.

So to find the equation of $y = f(x)$ after a stretch of scale factor 2 in the x-direction we replace x by $\tfrac{1}{2}x$ to obtain $y = f\left(\tfrac{1}{2}x\right)$.

K In general, in a stretch of scale factor k in the x-direction, x is replaced in the equation by $\frac{1}{k}x$. So the graph of $y = f(x)$ is mapped on to the graph of $y = f\left(\frac{1}{k}x\right)$.

It follows that, in a stretch of scale factor $\frac{1}{k}$ in the x-direction, x is replaced in the equation by $\left(\dfrac{1}{\left(\frac{1}{k}\right)}\right)x$, which is equivalent to kx.

K So the graph of $y = f(x)$ is mapped on to the graph of $y = f(kx)$ by a stretch of scale factor $\frac{1}{k}$ in the x-direction.

For example, $y = 2^x$ is changed to $y = 2^{4x}$ by replacing x by $4x$. So the graph of $y = 2^x$ is transformed on to the graph of $y = 2^{4x}$ by a stretch of scale factor $\frac{1}{4}$ in the x-direction.

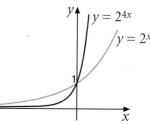

Now suppose that $y = f(x)$ is transformed by a stretch of scale factor 2 in the y-direction. Thus the y-coordinate of each point is doubled.

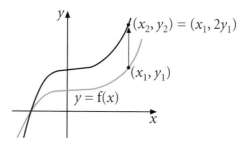

Let (x_1, y_1) be a point on $y = f(x)$ and let (x_2, y_2) be its image on the stretched curve.
Then $(x_2, y_2) = (x_1, 2y_1)$, giving

$$x_2 = x_1 \quad \text{which rearranges to} \quad x_1 = x_2$$
$$\text{and} \quad y_2 = 2y_1 \quad \text{which rearranges to} \quad y_1 = \tfrac{1}{2}y_2$$

We know that $y_1 = f(x_1)$, so it must be true that
$$\tfrac{1}{2}y_2 = f(x_2)$$
and so $\tfrac{1}{2}y = f(x)$ is the equation of the image.
This can be rearranged to give $y = 2f(x)$.

So to find the equation of $y = f(x)$ after a stretch of scale factor 2 in the y-direction we replace y by $\tfrac{1}{2}y$ to obtain $\tfrac{1}{2}y = f(x)$ or $y = 2f(x)$.

K In general, a stretch of scale factor k in the y-direction transforms an equation by replacing y with $\frac{1}{k}y$. So the graph of $y = f(x)$ is mapped on to the graph of $\frac{1}{k}y = f(x)$, which rearranges to $y = kf(x)$.

For example, a stretch of scale factor 3 in the y-direction replaces y with $\frac{1}{3}y$.
So stretching the graph of $y = \cos x$ by a factor of 3 in the y-direction transforms it on to the graph of $\frac{1}{3}y = \cos x$ or $y = 3\cos x$.

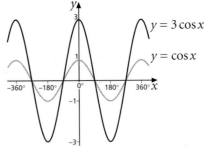

A1 Show that the image of $y = x^2 + 1$ after a stretch of factor $\frac{1}{2}$ in the x-direction is $y = 4x^2 + 1$.

A2 Show that a translation of $\begin{bmatrix} a \\ 0 \end{bmatrix}$ maps the graph of $y = f(x)$ on to the graph of $y = f(x - a)$.

A3 What is the image of $y = |x|$ after a translation of $\begin{bmatrix} 3 \\ 0 \end{bmatrix}$?

A4 What is the image of $y = f(x)$ after a translation of $\begin{bmatrix} 0 \\ a \end{bmatrix}$?

A5 Find the image of $y = 3^x$ after a translation of $\begin{bmatrix} 0 \\ 5 \end{bmatrix}$.

A6 What is the image of $y = f(x)$ after

    (a) reflection in the x-axis   **(b)** reflection in the y-axis

A7 What is the image of $y = x^2 + x$ after reflection in the y-axis?

Example 1

Find the equation of the image of $y = x^2 + 3x$ after a translation of $\begin{bmatrix} -2 \\ 0 \end{bmatrix}$.

Solution

For a translation of $\begin{bmatrix} -2 \\ 0 \end{bmatrix}$ we replace x by $x + 2$.

So $y = x^2 + 3x$ becomes $y = (x + 2)^2 + 3(x + 2) = x^2 + 4x + 4 + 3x + 6 = x^2 + 7x + 10$.

So the equation of the image is $y = x^2 + 7x + 10$.

Example 2

The function f is defined by $f(x) = x^3 - x$.

Show that $f(2x) = 8x^3 - 2x$.

The diagram shows the graph of $y = x^3 - x$.
Copy it and add the sketch of $y = 8x^3 - 2x$.

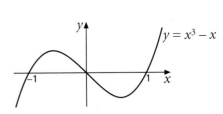

Solution

$f(2x) = (2x)^3 - (2x) = 8x^3 - 2x$ as required.

x is replaced by $2x$, so the transformation is a stretch of factor $\frac{1}{2}$ in the x-direction as shown.

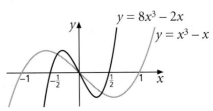

Exercise A (answers p 173)

1 Describe the transformation that transforms $y = f(x)$ on to each of these.

   (a) $y = 3f(x)$ (b) $y = f(3x)$ (c) $y = \frac{1}{3}f(x)$ (d) $y = f\left(\frac{1}{3}x\right)$

2 Find the image of $y = x^3$ after a stretch of factor 2 in the y-direction.

3 The function f is defined by $f(x) = x^2 - x$.
The diagram shows the graph of $y = x^2 - x$.

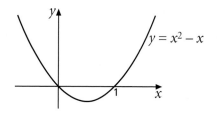

   (a) Show that $f(5x) = 25x^2 - 5x$.

   (b) Hence describe the transformation that maps $y = x^2 - x$ on to $y = 25x^2 - 5x$.

   (c) Sketch the graph of $y = 25x^2 - 5x$.

4 Describe the transformation that transforms $y = f(x)$ on to each of these.

   (a) $y = f(x) + 3$ (b) $y = f(x + 3)$ (c) $y = f(x) - 3$ (d) $y = f(x - 3)$

5 What transformation maps the graph of $y = \sin x$ on to the graph of $y = \sin(x + 90)$?

6 The function f is defined by $f(x) = x^2 + x$.

   (a) Show that $f(x - 4) = x^2 - 7x + 12$.

   (b) Hence describe the transformation that maps $y = x^2 + x$ on to $y = x^2 - 7x + 12$.

7 The function f is defined by $f(x) = x^3 - 4x$.
The diagram shows the graph of $y = x^3 - 4x$ and
its image after a translation in the x-direction.

Find the equation of the translated graph
in the form $y = ax^3 + bx^2 + cx + d$.

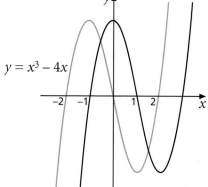

8 (a) Find the image of $y = x^4 + x^2$ after reflection in the y-axis.

   (b) Hence show that the graph of $y = x^4 + x^2$ has the y-axis as a line of symmetry.

9 (a) Find the image of $y = 5^x$ after   (i) a translation of $\begin{bmatrix} -1 \\ 0 \end{bmatrix}$

   (ii) a stretch of scale factor 5 in the y-direction

   (b) Show that these images are the same.

$y = f(x)$ is transformed on to …	by a …	that …
$y = f(x) + a$	translation of $\begin{bmatrix} 0 \\ a \end{bmatrix}$	replaces y by $(y - a)$
$y = f(x - a)$	translation of $\begin{bmatrix} a \\ 0 \end{bmatrix}$	replaces x by $(x - a)$
$y = f(x + a)$	translation of $\begin{bmatrix} -a \\ 0 \end{bmatrix}$	replaces x by $(x + a)$
$y = kf(x)$	stretch of factor k in the y-direction	replaces y by $\frac{1}{k}y$
$y = f\left(\frac{1}{k}x\right)$	stretch of factor k in the x-direction	replaces x by $\frac{1}{k}x$
$y = f(kx)$	stretch of factor $\frac{1}{k}$ in the x-direction	replaces x by kx
$y = -f(x)$	reflection in the x-axis	replaces y by $-y$
$y = f(-x)$	reflection in the y-axis	replaces x by $-x$

B Combining transformations (answers p 173)

B1 The graph of $y = x$ is first translated by $\begin{bmatrix} 0 \\ 2 \end{bmatrix}$ and then stretched by factor 3 in the y-direction. Sketch a graph of the final image and determine its equation.

B2 (a) The graph of $y = x$ is first translated by $\begin{bmatrix} 1 \\ 0 \end{bmatrix}$ and then reflected in the y-axis. Find the equation of the final image.

(b) The graph of $y = x$ is first reflected in the y-axis and then translated by $\begin{bmatrix} 1 \\ 0 \end{bmatrix}$. What is the equation of the final image?

B3 The graph of $y = x^2$ is first stretched by factor 3 in the y-direction and then translated by $\begin{bmatrix} 0 \\ -5 \end{bmatrix}$.
Show that the equation of the final image is $y = 3x^2 - 5$.

B4 Find the image of $y = x^2$ after each sequence of transformations.

(a) A translation of $\begin{bmatrix} 0 \\ 1 \end{bmatrix}$ followed by a reflection in the x-axis

(b) A translation of $\begin{bmatrix} 5 \\ 0 \end{bmatrix}$ followed by a stretch of factor 2 in the y-direction

(c) A translation of $\begin{bmatrix} -3 \\ 0 \end{bmatrix}$ followed by a translation of $\begin{bmatrix} 0 \\ -2 \end{bmatrix}$

(d) A stretch of factor 6 in the y-direction followed by reflection in the x-axis

B5 Describe a sequence of transformations that will map $y = x^2$ on to each of these.

(a) $y = \frac{1}{2}(x + 1)^2$ (b) $y = 7 - x^2$

B6 Which of these is the image of $y = f(x)$ after a translation of $\begin{bmatrix} 1 \\ 0 \end{bmatrix}$ followed by a stretch of factor 3 in the y-direction?

A $y = 3f(x + 1)$ **B** $y = 3f(x - 1)$ **C** $y = \frac{1}{3}f(x + 1)$ **D** $y = \frac{1}{3}f(x - 1)$

B7 Which of these is the image of $y = f(x)$ after a reflection in the y-axis followed by a stretch of factor 2 in the x-direction?

A $y = -2f(x)$ **B** $y = f\left(-\frac{1}{2}x\right)$ **C** $y = \frac{1}{2}f(-x)$ **D** $y = f(-2x)$

We can use algebra to determine equations of graphs after a sequence of transformations.

Example 3

The graph of $y = x^2$ is first stretched by factor 3 in the y-direction and then translated by $\begin{bmatrix} 2 \\ 0 \end{bmatrix}$. Find the equation of the final image and sketch its graph.

Solution

For the image of $y = x^2$ after a stretch of factor 3 in the y-direction, replace y by $\frac{1}{3}y$. So the image is $\frac{1}{3}y = x^2$, which is equivalent to $y = 3x^2$.

For the image of $y = 3x^2$ after a translation of $\begin{bmatrix} 2 \\ 0 \end{bmatrix}$, replace x by $(x - 2)$. So the final image is $y = 3(x - 2)^2$.

The sequence of graphs is

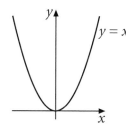 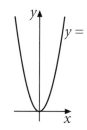 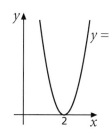

The final graph is the required result.

Example 4

The graph of $y = f(x)$ is first stretched by factor $\frac{1}{2}$ in the x-direction and then stretched by factor 5 in the y-direction.

Write down the coordinates of the image of the point $(4, 1)$ and find the equation of the final image.

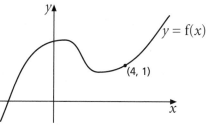

Solution

The first stretch multiplies the x-coordinate by $\frac{1}{2}$ and the second multiplies the y-coordinate by 5.

The coordinates of the image of the point are $\left(\frac{1}{2} \times 4, 5 \times 1\right) = (2, 5)$.

For the image of $y = f(x)$ after a stretch of factor $\frac{1}{2}$ in the x-direction, replace x by $2x$. So the image is $y = f(2x)$.

For the image of $y = f(2x)$ after a stretch of factor 5 in the y-direction, replace y by $\frac{1}{5}y$. So the final image is $\frac{1}{5}y = f(2x)$, which is equivalent to $y = 5f(2x)$.

Example 5

The diagram shows the graph of $y = 3 \times 2^{-x}$.

Describe a sequence of transformations by which this graph can be obtained from the graph of $y = 2^x$.

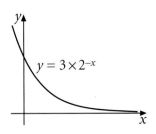

Solution

$y = 3 \times 2^{-x}$ is equivalent to $\frac{1}{3}y = 2^{-x}$.

From $y = 2^x$ we can obtain $\frac{1}{3}y = 2^{-x}$ by replacing y by $\frac{1}{3}y$ and x by $-x$.

So a sequence of transformations is a stretch of factor 3 in the y-direction followed by a reflection in the y-axis (or vice versa).

Example 6

The diagram shows the graph of $y = f(x)$, where f is defined for $0 \le x \le 3$.

Sketch the graph of $y = 2f(x + 1)$.

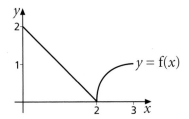

Solution

$y = 2f(x + 1)$ is equivalent to $\frac{1}{2}y = f(x + 1)$.

From $y = f(x)$ we can obtain $\frac{1}{2}y = f(x + 1)$ by replacing y by $\frac{1}{2}y$ and x by $(x + 1)$.

So a sequence of transformations is a stretch of factor 2 in the y-direction followed by a translation of $\begin{bmatrix} -1 \\ 0 \end{bmatrix}$ (or vice versa).

Hence a sequence of graphs is

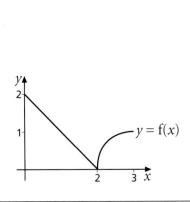

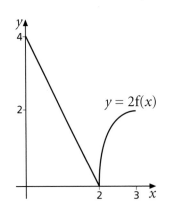

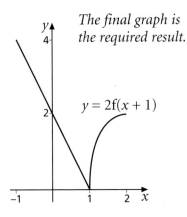

The final graph is the required result.

Exercise B (answers p 173)

1 The graph of $y = x^2$ is first translated by $\begin{bmatrix} 1 \\ 0 \end{bmatrix}$ and then translated by $\begin{bmatrix} 0 \\ 5 \end{bmatrix}$.
What is the equation of the final image?

2 The graph of $y = x^3$ is first stretched by a factor of 2 in the x-direction and then
translated by $\begin{bmatrix} 0 \\ -2 \end{bmatrix}$. Find the equation of the final image in the form $y = px^3 + q$,
where p and q are constants.

3 The graph of $y = |x|$ is first translated by $\begin{bmatrix} 3 \\ 0 \end{bmatrix}$ and then stretched by
a factor of 2 in the y-direction.

(a) What is the equation of the final image?

(b) Sketch the graph of the final image, showing clearly where
the graph meets each axis.

4 Describe a sequence of geometrical transformations that will map

(a) the graph of $y = x^2$ on to $y = \frac{1}{2}(x - 1)^2$

(b) the graph of $y = |x|$ on to $y = |3x| - 5$

(c) the graph of $y = 4^x$ on to $y = 4^{x+1} + 3$

5 State a sequence of geometrical transformations that will map $y = f(x)$ on to

(a) $y = 2f(4x)$ (b) $y = 3f(x + 1)$ (c) $y = \frac{1}{4}f(x - 2)$

6 The diagram shows the graph of $y = f(x)$
where the domain of f is $-2 \le x \le 2$.
The point $P\left(1, \sqrt{3}\right)$ lies on the curve.

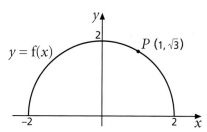

Sketch the graph of $y = \frac{1}{2}f(x - 2)$, showing clearly the coordinates of the image of P.

7 The function f is defined by $f(x) = x^2$.

(a) Show that $\frac{1}{2}f(x - 4) = \frac{1}{2}x^2 - 4x + 8$.

(b) Hence find a sequence of transformations that maps $y = x^2$ on to $y = \frac{1}{2}x^2 - 4x + 8$.

8 The diagram shows the graph of $y = f(x)$, $x \ge 0$.

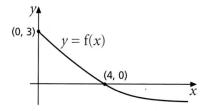

Sketch the graph of $y = 3f(4x)$, showing clearly where the graph meets each axis.

9 The diagram shows the graph of $y = f(x)$.

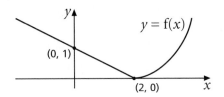

Sketch the graph of $y = 2f(x + 1)$, showing clearly the images of $(0, 1)$ and $(2, 0)$.

10 The diagram shows the graph of $y = f(x)$.

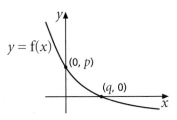

If the graph is transformed to give the graph of $y = f\left(\frac{1}{4}x\right) + 3$, what are the images of $(0, p)$ and $(q, 0)$?

11 The diagram shows the graph of $y = f(x)$.

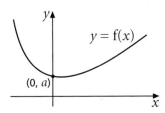

Sketch the graph of $y = 2f(-x)$, showing the coordinates of the image of $(0, a)$.

C Order of transformations (answers p 174)

Sometimes the order that transformations are applied in matters and sometimes it does not.

C1 (a) The graph of $y = x$ is first translated by $\begin{bmatrix} 0 \\ 3 \end{bmatrix}$ and then stretched by factor 2 in the y-direction.
Sketch a graph of the final image and find its equation.

 (b) Now $y = x$ is transformed using the same transformations but in the opposite order, first stretching by factor 2 in the y-direction and then translating by $\begin{bmatrix} 0 \\ 3 \end{bmatrix}$.
Find the equation of the final image.

 (c) Are the final images the same?

C2 (a) The graph of $y = x^2$ is first translated by $\begin{bmatrix} 1 \\ 0 \end{bmatrix}$ and then reflected in the x-axis.
Find the equation of the final image.

 (b) Now $y = x^2$ is transformed using the same transformations but in the opposite order, first reflecting in the x-axis and then translating by $\begin{bmatrix} 1 \\ 0 \end{bmatrix}$.
Find the equation of the final image.

 (c) Can you explain why both these final images are the same?

We can use algebra to decide when order matters in applying a sequence of transformations to $y = f(x)$.

Consider the transformations • a translation by $\begin{bmatrix} 0 \\ 2 \end{bmatrix}$

• a stretch of factor 3 in the y-direction.

When $y = f(x)$ is translated by $\begin{bmatrix} 0 \\ 2 \end{bmatrix}$ we replace y by $y - 2$ to obtain $y - 2 = f(x)$, which is $y = f(x) + 2$.

If we now stretch by a factor of 3 in the y-direction we replace y by $\frac{1}{3}y$ to obtain $\frac{1}{3}y = f(x) + 2$, which is $y = 3f(x) + 6$.

Now apply the transformations in the opposite order.

When $y = f(x)$ is stretched by a factor of 3 in the y-direction we replace y by $\frac{1}{3}y$ to obtain $\frac{1}{3}y = f(x)$, which is $y = 3f(x)$.

If we now translate by $\begin{bmatrix} 0 \\ 2 \end{bmatrix}$ we replace y by $y - 2$ to obtain $y - 2 = 3f(x)$, which is $y = 3f(x) + 2$.

The equations of the final images are different: the order matters here.

> **K** When applying two transformations, the order does not matter if one transformation involves replacing x and the other involves replacing y.
>
> If both transformations involve replacing x (or y), then the order could matter.

Example 7

Find a sequence of transformations that maps the graph of $y = f(x)$ on to $y = 2f(x) - 5$.

Solution

$y = 2f(x)$ *is equivalent to* $\frac{1}{2}y = f(x)$.

First, $y = f(x)$ can be transformed to $y = 2f(x)$ by a stretch of factor 2 in the y-direction.

$y = 2f(x) - 5$ *is equivalent to* $y + 5 = 2f(x)$.

Then $y = 2f(x)$ can be transformed to $y = 2f(x) - 5$ by a translation of $\begin{bmatrix} 0 \\ -5 \end{bmatrix}$.

So a sequence is a stretch of factor 2 in the y-direction followed by a translation of $\begin{bmatrix} 0 \\ -5 \end{bmatrix}$.

Exercise C (answers p 174)

1 (a) The graph of $y = x^2$ is first translated by $\begin{bmatrix} 0 \\ -2 \end{bmatrix}$ and then stretched by factor 5 in the y-direction.

Find the equation of the final graph.

(b) Now $y = x^2$ is transformed using the same transformations but in the opposite order, first stretching by factor 5 in the y-direction and then translating by $\begin{bmatrix} 0 \\ -2 \end{bmatrix}$.

Find the equation of the final graph. What do you notice?

2 Find the image of $y = f(x)$ after each sequence of transformations.

(a) A stretch of factor $\frac{1}{2}$ in the y-direction followed by a translation of $\begin{bmatrix} 0 \\ -3 \end{bmatrix}$

(b) A translation of $\begin{bmatrix} 0 \\ 4 \end{bmatrix}$ followed by a reflection in the x-axis

(c) A translation of $\begin{bmatrix} 0 \\ -4 \end{bmatrix}$ followed by a stretch of factor $\frac{1}{4}$ in the y-direction

3 The graph of $y = f(x)$ is translated by $\begin{bmatrix} 3 \\ 0 \end{bmatrix}$ and stretched by factor 2 in the y-direction. Show that using these transformations in any order will result in the same final image.

4 (a) Find the image of $y = f(x)$ after a stretch of factor k in the y-direction followed by a translation of $\begin{bmatrix} 0 \\ a \end{bmatrix}$.

(b) Use your result to describe a sequence of transformations that will map $y = f(x)$ on to $y = 3f(x) - 5$.

5 Find a sequence of transformations that will map $y = f(x)$ on to

(a) $y = \frac{1}{3}f(x) + 1$ **(b)** $y = -f(x) + 5$ **(c)** $y = 3f(x) - 6$

6 (a) What is the image of $y = f(x)$ after

(i) a stretch of scale factor 2 in the y-direction followed by a translation of $\begin{bmatrix} 0 \\ 10 \end{bmatrix}$

(ii) a translation of $\begin{bmatrix} 0 \\ 5 \end{bmatrix}$ followed by a stretch of scale factor 2 in the y-direction

(b) What can you deduce from your answers to part (a)?

7 (a) What is the image of $y = f(x)$ after

(i) a reflection in the x-axis followed by a translation of $\begin{bmatrix} 0 \\ -2 \end{bmatrix}$

(ii) a translation of $\begin{bmatrix} 0 \\ 2 \end{bmatrix}$ followed by a reflection in the x-axis

(b) What can you deduce from your answers to part (a)?

Key points

- The rules for the basic transformations are on page 60.

- A graph can be transformed using a combination of basic transformations. You can determine the equation of the final image by applying the rules for the basic transformations in order.

  For example, to find the result of transforming $y = f(x)$ by a translation of $\begin{bmatrix} 0 \\ 5 \end{bmatrix}$ followed by a stretch of factor $\frac{1}{2}$ in the x-direction, first replace y by $y - 5$ and then replace x by $2x$ to obtain $y - 5 = f(2x)$ or $y = f(2x) + 5$. (pp 61–62)

- When applying two transformations, the order does not matter if one transformation involves replacing x and the other involves replacing y. If both transformations involve replacing x (or y), then the order could matter. (pp 64–65)

Test yourself (answers p 175)

1 The diagram shows a sketch of $y = \frac{1}{2}(x + 1)^3$.

Describe fully a sequence of geometrical transformations that would map the graph of $y = x^3$ on to the graph of $y = \frac{1}{2}(x + 1)^3$.

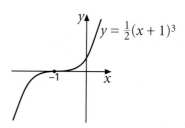

2 A sketch of the graph of $y = f(x)$ is shown. The point $P\ (3, 2)$ lies on the graph.

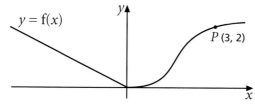

(a) Describe fully a sequence of transformations that maps the graph of $y = f(x)$ on to the graph of $y = 4f(x - 3)$.

(b) Sketch the graph of $y = 4f(x - 3)$. Label the image of the point P, giving its coordinates.

3 Describe a sequence of transformations by which the graph of $y = 4 \times 5^{-x}$ can be obtained from the graph of $y = 5^x$.

4 The diagram shows the sketch of a curve with equation $y = f(x)$. The curve meets the axes at the points $(a, 0)$ and $(0, b)$.

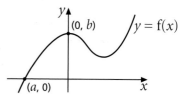

(a) Describe a sequence of transformations by which the graph of $y = f(x)$ can mapped on to the graph of $y = 2f\left(\frac{1}{3}x\right)$.

(b) Sketch the curve with equation $y = 2f\left(\frac{1}{3}x\right)$, showing clearly where the curve meets the coordinate axes.

5 The graph of $y = x^2$ is first translated by $\begin{bmatrix} 0 \\ -3 \end{bmatrix}$ and then stretched by factor 2 in the y-direction. Find the equation of the final image.

6 Determine a sequence of transformations that will map $y = f(x)$ on to

(a) $y = 3f(x) + 4$ (b) $y = f(-x) - 7$

7 Find the image of $y = f(x)$ after reflection in the x-axis followed by a translation of $\begin{bmatrix} 0 \\ 6 \end{bmatrix}$.

5 Trigonometry

In this chapter you will learn about
- the inverse functions of cosine, sine and tangent
- the reciprocal functions secant, cosecant and cotangent
- transformations of the above functions
- identities and equations involving the above functions

A Inverse circular functions (answers p 176)

This is part of the graph of $y = \sin x$, where x is measured in radians.

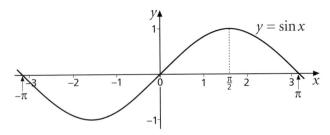

A1 Using the same scale on each axis, make a sketch copy of the graph above.
Draw the line $y = x$ on your sketch.
Draw the reflection of $y = \sin x$ in the line $y = x$ on your sketch.
Mark carefully the position of the reflection of the point $\left(\frac{\pi}{2}, 1\right)$.

A2 You saw in chapter 2 that reflecting the graph of a function in the line $y = x$ may give the graph of an inverse function.
Does it give an inverse function in this case? If not, why not?

In order that the reflection of $y = \sin x$ in the line $y = x$ may represent a function, we need to restrict the domain of $y = \sin x$, so that $y = \sin x$ is a one–one function.

A3 Which of these are one–one functions?
$$f(x) = \sin x, \ -\frac{\pi}{2} \le x \le \frac{\pi}{2} \qquad g(x) = \sin x, \ 0 \le x \le \pi \qquad h(x) = \sin x, \ \frac{\pi}{2} \le x \le \frac{3\pi}{2}$$
How we restrict the domain of $y = \sin x$ so it has an inverse is to some extent arbitrary, but the convention is that we restrict x to the interval $-\frac{\pi}{2} \le x \le \frac{\pi}{2}$.
The inverse function of $\sin x$ is denoted by $y = \arcsin x$.
The range of $\arcsin x$ is the restricted domain of $\sin x$, that is $-\frac{\pi}{2} \le \arcsin x \le \frac{\pi}{2}$.
We call the range of $\arcsin x$ the **principal values** of $\arcsin x$.
So, for example, $\arcsin 1 = \frac{\pi}{2}$ $\left(\frac{\pi}{2}\right.$ is the principal value for the angle whose sine is $1\left.\right)$.

A4 What is the domain of $\arcsin x$?

The graph of $y = \arcsin x$ is shown on the right.

We can also write $\sin^{-1} x$ or $\text{a}\sin x$ for $\arcsin x$.

(Note that $\sin^{-1} x$ never means $\dfrac{1}{\sin x}$.)

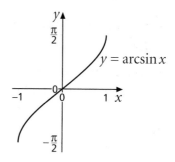

K   The domain of $y = \arcsin x$ is $-1 \leq x \leq 1$.

The range of $y = \arcsin x$ is $-\dfrac{\pi}{2} \leq y \leq \dfrac{\pi}{2}$.

A5 Using degrees, $\arcsin 1 = 90°$. In degrees, write down the value of

(a) $\arcsin \frac{1}{2}$ (b) $\arcsin\left(\dfrac{1}{\sqrt{2}}\right)$ (c) $\arcsin 0$ (d) $\arcsin(-1)$

A6 In order for $\tan x$ to have an inverse, we restrict its domain to $-\dfrac{\pi}{2} \leq x \leq \dfrac{\pi}{2}$.

(a) With this domain, what is the range of $\tan x$?

(b) Using equal scales on the x- and y-axes, sketch the graph of $y = \tan x$ with domain as stated above.

(c) On the same graph, sketch and label the inverse function of $\tan x$, which is $y = \arctan x$.

(d) Write down the domain and range of $y = \arctan x$.

Example 1

Write down, in degrees, the value of

(a) $\arcsin\left(-\dfrac{1}{\sqrt{2}}\right)$ (b) $\arctan\left(-\dfrac{1}{\sqrt{3}}\right)$

Solution

You need to know, or to be able quickly to work out, the sine, cosine and tangent of $0°$, $30°$, $45°$, $60°$ and $90°$.

(a) $\sin 45° = \dfrac{1}{\sqrt{2}}$, so $\sin(-45°) = -\dfrac{1}{\sqrt{2}}$.

arcsin x is between $-90°$ and $90°$.   Hence $\arcsin\left(-\dfrac{1}{\sqrt{2}}\right) = -45°$.

(b) $\tan 30° = \dfrac{1}{\sqrt{3}}$, so $\tan(-30°) = -\dfrac{1}{\sqrt{3}}$.

arctan x is between $-90°$ and $90°$.   Hence $\arctan\left(-\dfrac{1}{\sqrt{3}}\right) = -30°$.

Exercise A (answers p 176)

1 In order for $\cos x$ to have an inverse, we restrict its domain to $0 \leq x \leq \pi$.

    (a) Using equal scales on the x- and y-axes, sketch the graph of $y = \cos x$ with the above domain.

    (b) On the same graph, sketch the inverse function of $\cos x$, $y = \arccos x$.

    (c) Write down the domain and range of $y = \arccos x$.

2 (a) Write down the solution in degrees, between $0°$ and $90°$, of $\cos\theta° = \frac{1}{2}$.

    (b) Hence write down the two solutions to $\cos\theta° = -\frac{1}{2}$ between $0°$ and $360°$.

    (c) What is the principal value of $\arccos\left(-\frac{1}{2}\right)$ in degrees?

3 Write down the value, in degrees, of each of the following

    (a) $\arccos(-1)$ **(b)** $\arccos\left(\frac{\sqrt{3}}{2}\right)$ **(c)** $\arccos 0$ **(d)** $\arctan\left(-\sqrt{3}\right)$

4 Write down, in radians, the value of each of these.

    (a) $\arccos 1$ **(b)** $\arccos\left(-\frac{1}{2}\right)$ **(c)** $\arctan(-1)$ **(d)** $\arcsin\left(-\frac{\sqrt{3}}{2}\right)$

    (e) $\arccos 0$ **(f)** $\arctan 0$ **(g)** $\arccos\left(-\frac{1}{\sqrt{2}}\right)$ **(h)** $\arcsin\left(-\frac{1}{2}\right)$

5 Solve these where possible.

    (a) $\arcsin x = \frac{\pi}{4}$ **(b)** $\arcsin x = \frac{\pi}{6}$ **(c)** $\arcsin x = -\frac{\pi}{4}$ **(d)** $\arcsin x = \pi$

6 (a) Work these out.

    (i) $\sin\left(\arcsin\frac{1}{2}\right)$ **(ii)** $\tan(\arctan 1)$ **(iii)** $\cos\left(\arccos\frac{\sqrt{3}}{2}\right)$ **(iv)** $\sin\left(\arcsin-\frac{1}{2}\right)$

    (b) Is it always true that $\sin(\arcsin x) = x$?

7 (a) Work these out.

    (i) $\arcsin\left(\sin\frac{\pi}{6}\right)$ **(ii)** $\arcsin\left(\sin\frac{5\pi}{6}\right)$ **(iii)** $\arcsin\left(\sin\frac{7\pi}{6}\right)$ **(iv)** $\arcsin\left(\sin\frac{13\pi}{6}\right)$

    (b) Is it always true that $\arcsin(\sin x) = x$?

8 (a) Which of the following domains ensures that the function $f(x) = \sin 2x$ is one–one?

    A $-\pi \leq x \leq \pi$ **B** $-\frac{\pi}{2} \leq x \leq \frac{\pi}{2}$ **C** $-\frac{\pi}{4} \leq x \leq \frac{\pi}{4}$

    (b) With the chosen domain, what is the range of $f(x)$?

    (c) Write down the domain and range of the inverse function, $f^{-1}(x)$.

*9 (a) Using your answers to question 8, suggest a possible inverse function for $y = \sin 2x$.

    (b) Check your suggestion, using a graph plotter, by

        (i) reflecting the graph of $y = \sin 2x$ in the line $y = x$

        (ii) plotting the graph of your suggestion and checking that it coincides with (i).

B Sec, cosec and cot

The most useful trigonometrical functions are cosine, sine and tangent. However there are three other closely related functions – secant, cosecant and cotangent (usually abbreviated to sec, cosec and cot).

These are defined as

$$\sec x = \frac{1}{\cos x} \qquad \operatorname{cosec} x = \frac{1}{\sin x} \qquad \cot x = \frac{1}{\tan x}$$

B1 (a) When $x = \frac{\pi}{4}$, what is the value of $\cos x$?

(b) What is the value of $\sec x$ when $x = \frac{\pi}{4}$?

B2 What is the value of $\sec x$ when

(a) $x = \frac{\pi}{6}$ (b) $x = \frac{\pi}{3}$ (c) $x = 0$ (d) $x = -\frac{\pi}{4}$

B3 Write down the value of $\operatorname{cosec} x$ when

(a) $x = \frac{\pi}{2}$ (b) $x = -\frac{\pi}{2}$ (c) $x = \frac{\pi}{4}$ (d) $x = -\frac{\pi}{6}$

B4 Evaluate these.

(a) $\cot \frac{\pi}{4}$ (b) $\cot \frac{\pi}{6}$ (c) $\cot \left(-\frac{\pi}{3}\right)$ (d) $\cot \frac{5\pi}{4}$

B5 (a) What happens when you try to evaluate $\operatorname{cosec} 0$?

(b) Write down two other values of x for which $\operatorname{cosec} x$ is not defined.

(c) Describe accurately the complete set of values of x for which $\operatorname{cosec} x$ is not defined.

B6 (a) Write down three values of x for which $\sec x$ is not defined.

(b) Describe clearly the complete set of values of x for which $\sec x$ is not defined.

The diagram shows part of the graph of $\cos x$.
Part of the graph of $\sec x$ is also shown.

B7 (a) Copy the diagram and complete the graph of $y = \sec x$ for $-\pi \leq x \leq 2\pi$.

(b) Describe clearly the domain of $\sec x$ (you will need to use your answer to B6).

(c) Write down an expression for the range of $y = \sec x$.

(d) What is the period of $\sec x$?

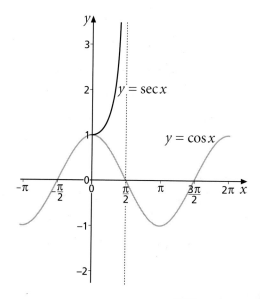

B8 (a) Sketch the graph of $\sin x$.

(b) On your diagram, sketch the graph of $\operatorname{cosec} x$.

(c) Write down expressions for the domain and range of $\operatorname{cosec} x$.

(d) What is the period of $\operatorname{cosec} x$?

The cotangent of x, $\cot x$, is defined as $\dfrac{1}{\tan x}$.

Using the relationship between $\sin x$ and $\cos x$, we can establish a relationship between $\sec x$ and $\tan x$.

$$\sin^2 x + \cos^2 x = 1$$

Dividing by $\cos^2 x$ we obtain
$$\frac{\sin^2 x}{\cos^2 x} + 1 = \frac{1}{\cos^2 x}$$

$$\Rightarrow \quad \tan^2 x + 1 = \sec^2 x$$

B9 Divide the relationship $\sin^2 x + \cos^2 x = 1$ by $\sin^2 x$ to obtain a relationship between $\operatorname{cosec} x$ and $\cot x$.

Example 2

Show that $\dfrac{\tan\theta}{1 + \tan^2\theta} = \sin\theta\cos\theta$.

Solution

$$\frac{\tan\theta}{1+\tan^2\theta} = \frac{\tan\theta}{\sec^2\theta} = \frac{\tan\theta}{\dfrac{1}{\cos^2 x}} = \tan\theta\times\cos^2\theta = \frac{\sin\theta}{\cos\theta}\times\cos^2\theta = \sin\theta\cos\theta$$

Example 3

Given that $\tan\theta = 2$, find exact values for **(a)** $\cot\theta$ **(b)** $\sec\theta$ **(c)** $\operatorname{cosec}\theta$

Solution

(a) $\cot\theta = \dfrac{1}{\tan\theta} = \frac{1}{2}$

One possible pair of values for $\sin\theta$ and $\cos\theta$ is $\sin\theta = \dfrac{2}{\sqrt{5}}$ and $\cos\theta = \dfrac{1}{\sqrt{5}}$.

However, both sine and cosine could be negative $\left(\text{for an angle between } \pi \text{ and } \dfrac{3\pi}{2}\right).$

sine positive	all positive
tangent positive	cosine positive

$\dfrac{\pi}{2}$ (top), π (left), 0 (right), $\dfrac{3\pi}{2}$ (bottom)

In this case, $\sin\theta = -\dfrac{2}{\sqrt{5}}$ and $\cos\theta = -\dfrac{1}{\sqrt{5}}$. Hence

(b) $\sec\theta = \dfrac{1}{\cos\theta} = \dfrac{1}{\dfrac{1}{\sqrt{5}}}$ or $\dfrac{1}{-\dfrac{1}{\sqrt{5}}} = \sqrt{5}$ or $-\sqrt{5}$

(c) $\operatorname{cosec}\theta = \dfrac{1}{\sin\theta} = \dfrac{1}{\dfrac{2}{\sqrt{5}}}$ or $\dfrac{1}{-\dfrac{2}{\sqrt{5}}} = \dfrac{\sqrt{5}}{2}$ or $-\dfrac{\sqrt{5}}{2}$

Example 4

Simplify $\operatorname{cosec}\left(\dfrac{\pi}{2} - x\right)$.

Solution

$$\operatorname{cosec}\left(\frac{\pi}{2} - x\right) = \frac{1}{\sin\left(\dfrac{\pi}{2} - x\right)} = \frac{1}{\cos x} = \sec x$$

K

$$\sec x = \frac{1}{\cos x} \qquad\qquad \operatorname{cosec} x = \frac{1}{\sin x} \qquad\qquad \cot x = \frac{1}{\tan x}$$

$$\sec^2 x = 1 + \tan^2 x \qquad\qquad \operatorname{cosec}^2 x = 1 + \cot^2 x$$

Exercise B (answers p 177)

1 (a) Sketch the graph of $\tan x$ for $-2\pi < x < 2\pi$.

   (b) On your diagram, sketch also the graph of $\cot x$.

   (c) Give the domain and range of $\cot x$, clearly stating any values that are excluded.

   (d) What is the period of $\cot x$?

2 Find the exact values of

   (a) $\operatorname{cosec} \dfrac{\pi}{3}$ (b) $\cot \dfrac{\pi}{3}$ (c) $\sec\left(-\dfrac{\pi}{3}\right)$ (d) $\operatorname{cosec}\left(-\dfrac{\pi}{3}\right)$

3 Simplify these.

   (a) $\operatorname{cosec} x \tan x$ (b) $\tan x \cot x$ (c) $\sin x \operatorname{cosec} x$ (d) $\sec x \sin x$

4 Simplify these.

   (a) $\dfrac{\cos\theta}{\sin\theta}$ (b) $\dfrac{\operatorname{cosec}\theta}{\sec\theta}$ (c) $\sec\left(\dfrac{\pi}{2} - x\right)$ (d) $\cot(\pi + x)$

5 Simplify the following.

   (a) $\dfrac{\sin x}{1 + \cot^2 x}$ (b) $\cos x\sqrt{1 + \tan^2 x}$

6 Prove the following identities.

   (a) $\tan\theta + \cot\theta = \sec\theta \operatorname{cosec}\theta$ (b) $\cot\theta \sec\theta = \operatorname{cosec}\theta$

7 Simplify these.

   (a) $\operatorname{cosec}^2 x - 1$ (b) $(\sec x + 1)(\sec x - 1)$ (c) $\dfrac{\tan x}{1 - \sec^2 x}$

8 Show that $\operatorname{cosec} x + \cot x = \dfrac{1}{\operatorname{cosec} x - \cot x}$ $(\operatorname{cosec} x \neq \cot x)$.

9 Given that $\sin\theta = \frac{4}{5}$, and that θ is in the first quadrant, find exact values of

   (a) $\cos\theta$ (b) $\tan\theta$ (c) $\sec\theta$ (d) $\operatorname{cosec}\theta$

10 Given that $\dfrac{\pi}{2} \leq \theta \leq \pi$ and $\cot\theta = -2\frac{2}{5}$, find exact values of

   (a) $\tan\theta$ (b) $\cos\theta$ (c) $\sin\theta$ (d) $\operatorname{cosec}\theta$

11 Given that $\tan\theta = 1$, what are the possible values of

   (a) $\cot\theta$ (b) $\sin\theta$ (c) $\cos\theta$ (d) $\operatorname{cosec}\theta$

C Solving equations (answers p 178)

When solving equations involving $\sec x$, $\operatorname{cosec} x$ or $\cot x$, it is often best to substitute so that the equation involves $\sin x$, $\cos x$ or $\tan x$ instead.

You may also be able to make use of the relations $\sec^2 x = 1 + \tan^2 x$ and $\operatorname{cosec}^2 x = 1 + \cot^2 x$.

Example 5

Solve the equation $\sec x = 2$ for $0 \le x \le 2\pi$.

Solution

$$\sec x = 2$$

Substitute for $\sec x$.

$$\frac{1}{\cos x} = 2$$

$$\Rightarrow \quad \cos x = \tfrac{1}{2}$$

$$\Rightarrow \quad x = \frac{\pi}{3} \text{ or } 2\pi - \frac{\pi}{3} = \frac{5\pi}{3}$$

Example 6

Solve $\cot^2 \theta° = 3$, where $-180° \le \theta° \le 180°$.

Solution

$$\cot^2 \theta° = 3$$

Substitute $\dfrac{1}{\tan \theta°}$ for $\cot \theta°$.

$$\frac{1}{\tan^2 \theta°} = 3$$

$$\Rightarrow \quad \tan^2 \theta° = \tfrac{1}{3}$$

$$\Rightarrow \quad \tan \theta° = \pm \frac{1}{\sqrt{3}}$$

If $\tan \theta° = \dfrac{1}{\sqrt{3}}$, then $\theta° = 30°$ or $-150°$; if $\tan \theta° = -\dfrac{1}{\sqrt{3}}$, then $\theta° = -30°$ or $150°$.

Hence $\theta° = -150°, -30°, 30°$ or $150°$.

Example 7

Solve the equation $\tan^2 x + \sec x = 1$ for $0 \le x \le 2\pi$.

Solution

$$\tan^2 x + \sec x = 1$$

You need to get a quadratic in a single variable, so substitute $\tan^2 x = \sec^2 x - 1$.

$$\sec^2 x - 1 + \sec x = 1$$

$$\Rightarrow \quad \sec^2 x + \sec x - 2 = 0$$

Factorise.

$$(\sec x + 2)(\sec x - 1) = 0$$

$$\Rightarrow \quad \sec x = -2 \text{ or } \sec x = 1$$

$$\Rightarrow \quad \cos x = -\tfrac{1}{2} \text{ or } \cos x = 1$$

$$\Rightarrow \quad x = \frac{2\pi}{3} \text{ or } \frac{4\pi}{3} \text{ or } x = 0 \text{ or } 2\pi$$

C1 Solve $\tan^2 x + \sec x = 1$ by substituting $\tan x = \dfrac{\sin x}{\cos x}$ and $\sec x = \dfrac{1}{\cos x}$ as the first step, and then solving a quadratic in $\cos x$.

Check your answer is identical to that in example 7.

Exercise C (answers p 178)

1 (a) Solve $\sec \theta° = \sqrt{2}$, giving exact answers within the range 0° to 360°.

(b) Hence solve $\sec^2 \theta° = 2$ $(0° \le \theta° \le 360°)$.

2 Solve $\cot \theta° = 3$, giving answers in the range −180° to 180° to the nearest degree.

3 Solve for $0 \le x \le 2\pi$

(a) $\cot x = 1$ **(b)** $\operatorname{cosec} x = 1$ **(c)** $\operatorname{cosec}^2 x = 1$ **(d)** $3 \operatorname{cosec}^2 x = 4$

4 Solve, giving answers in the range −π to π to two decimal places,

(a) $\sec x = 3$ **(b)** $\sec^2 x = 9$ **(c)** $\operatorname{cosec} x = 3$ **(d)** $\cot x = \frac{1}{2}$

5 (a) By replacing $\sec^2 x$ by $1 + \tan^2 x$, show that the equation

$$\sec^2 x + \tan x = 3$$

is equivalent to the equation

$$\tan^2 x + \tan x - 2 = 0$$

(b) Factorise the left-hand side of this equation.

(c) Solve the equation to find all the values of x between −π and π.

6 By substituting for $\cot^2 \theta$ in terms of $\operatorname{cosec}^2 \theta$, solve $2 \cot^2 \theta + \operatorname{cosec} \theta + 1 = 0$ for $-\pi \le \theta \le \pi$.

7 Solve $2 \tan^2 x - 7 \sec x + 5 = 0$, giving answers between 0 and 2π to one decimal place.

8 Solve $3 \cot x + 2 \tan x = 5$. Give answers between 0° and 90° to one decimal place.

9 Solve the following equations for $0° \le \theta° \le 360°$.
Give your answers to two decimal places.

(a) $2 \sec^2 \theta° = 9 \tan \theta° + 7$ **(b)** $\operatorname{cosec}^2 \theta° = 3 \cot \theta° + 5$

10 Solve $\sec x = \tan x$ for $0 \le x \le 2\pi$.
Explain your result with the aid of a graph.

11 The diagram shows the graphs of $y = \sin x$ and $y = \cot x$ $(0 \le x \le 2\pi)$.

Find the coordinates of the points A and B, to two decimal places.

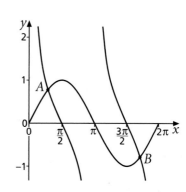

D Transforming graphs (answers p 178)

In chapter 4 you saw how transforming the graph of a function affects its equation.

The diagrams below show how $y = \sin x$ (shown in grey) is changed after the transformations shown.

$y = a\sin x$:
stretch in the y-direction, scale factor a

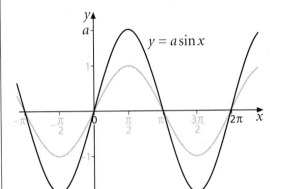

$y = \sin bx$:
stretch in the x-direction, scale factor $\dfrac{1}{b}$

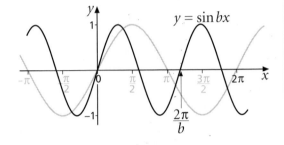

$y = \sin(x + c)$: a translation by $\begin{bmatrix} -c \\ 0 \end{bmatrix}$

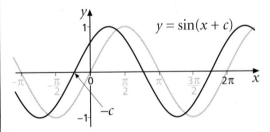

$y = (\sin x) + d$: a translation by $\begin{bmatrix} 0 \\ d \end{bmatrix}$

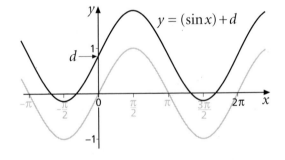

$y = -\sin x$:
reflection in the x-axis

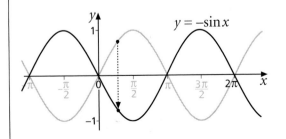

$y = \sin(-x)$:
reflection in the y-axis

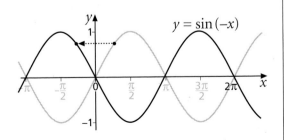

D1 Without doing any plotting, state the two transformations that will transform $y = \sin x$ on to each of these.

(a) $y = \frac{1}{3} \sin(-x)$

(b) $y = \sin(2x) + 3$

(c) $y = 3 - \sin x$ (note: $3 - \sin x = -(\sin x) + 3$)

D2 (a) What is the equation of the image of $y = \cos x$ after each of the following?

(i) A translation by $\begin{bmatrix} 0 \\ 2 \end{bmatrix}$, followed by a stretch in the x-direction, scale factor 4

(ii) A stretch in the y-direction, scale factor 2, followed by a translation by $\begin{bmatrix} 1 \\ 0 \end{bmatrix}$

(iii) A stretch in the x-direction, scale factor 3, followed by a stretch in the y-direction, scale factor 2

(b) For each part of (a), sketch $y = \cos x$ and the result of the transformations.

D3 For each of the graphs below,

(i) state two transformations that will transform $y = \sin x$ (shown in grey on each diagram) on to the given graph

(ii) hence write down the equation of the graph

(a)

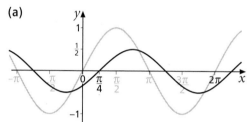

(b)

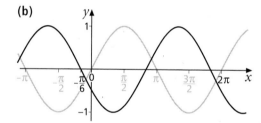

(c)

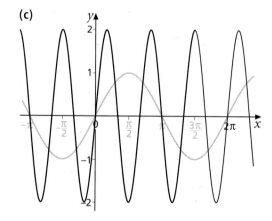

(d)

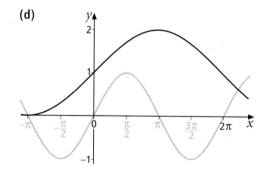

Example 8

Sketch the graph of $y = \cos\theta°$ $(-360° \leq \theta° \leq 360°)$, and its image after a stretch in the θ-direction, scale factor 2, followed by a stretch in the y-direction, scale factor 1.5.

What is the equation of the image?

Solution

The graph of $y = \cos\theta°$ is shown in grey; shown dotted after the stretch in the θ-direction, and solid after the stretch in the y-direction.

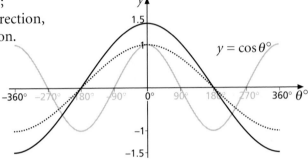

The graph of $y = \cos\theta°$ becomes $y = \cos\frac{1}{2}\theta°$ after a stretch in the θ-direction, scale factor 2.

The graph of $y = \cos\frac{1}{2}\theta°$ becomes $y = 1.5\cos\frac{1}{2}\theta°$ after a stretch in the y-direction, scale factor 1.5.

The final graph has the equation $y = 1.5\cos\frac{1}{2}\theta°$.

It is sensible to now check this by drawing the graph of $y = 1.5\cos\frac{1}{2}\theta°$ on a graph plotter and seeing whether it agrees with the graph you have sketched.

Example 9

State two transformations which, when applied to the graph of $y = \tan x$, will give the graph of $y = \tan 2x - 1$. Hence sketch the graph of $y = \tan 2x - 1$, where $-2\pi \leq x \leq 2\pi$.

Solution

The graph of $y = \tan 2x - 1$ can be obtained from that of $y = \tan x$ by first applying a stretch in the x-direction, scale factor $\frac{1}{2}$ (giving $y = \tan 2x$) and then by applying the translation $\begin{bmatrix} 0 \\ -1 \end{bmatrix}$ to $y = \tan 2x$ (giving $y = \tan 2x - 1$).

Hence the graph is

It is sensible to check this on a graph plotter.

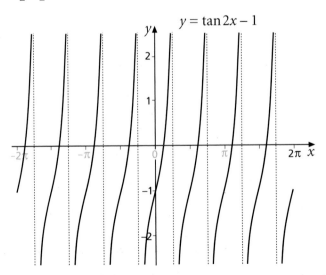

Example 10

Sketch the graph of $y = \arccos 2x$. State its domain and range.

Solution

Start by sketching the graph of $y = \arccos x$.

The graph of $y = \arccos x$ is shown dotted on the right.

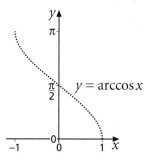

We obtain the graph of $y = \arccos 2x$ from that of $y = \arccos x$ by replacing x by $2x$, that is by applying a stretch in the x-direction with scale factor $\frac{1}{2}$.

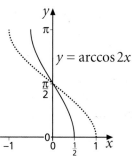

Transforming $y = \arccos x$ in this way we obtain the graph of $y = \arccos 2x$ (shown solid).

The domain is $-\frac{1}{2} \le x \le \frac{1}{2}$, and the range is $0 \le y \le \pi$.

Exercise D (answers p 179)

1 Each of these graphs shows $y = \cos x$ after a single transformation. Identify the transformation and thus the equation of each graph.

(a)

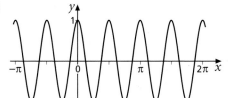

(b)

(c)

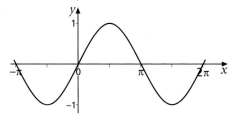

(d)

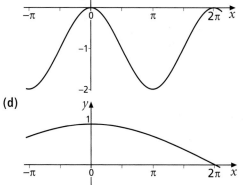

2 (a) Find the equation of the resulting graph when $y = \tan x$ is transformed by

  (i) a stretch, scale factor 2, in the x-direction, followed by a translation by $\begin{bmatrix} 0 \\ -1 \end{bmatrix}$

  (ii) a reflection in the y-axis, followed by a stretch, factor $\frac{1}{2}$, in the x-direction

  (iii) a translation by $\begin{bmatrix} 1 \\ 0 \end{bmatrix}$ followed by reflection in the x-axis

  (iv) a translation by $\begin{bmatrix} 0 \\ 1 \end{bmatrix}$ followed by reflection in the y-axis

  (b) For each part of (a), sketch $y = \tan x$ and the result of the transformations.

3 (a) The point $A\left(\frac{\pi}{6}, \frac{1}{2}\right)$ lies on the graph of $y = \sin x$.

What are the coordinates of the image of A after a stretch, factor 2,

in the y-direction, followed by a translation by $\begin{bmatrix} \frac{\pi}{4} \\ 0 \end{bmatrix}$?

(b) The graph of $y = \sin x$ is transformed by the pair of transformations in (a). What is the equation of the transformed graph?

(c) Check that the coordinates of the image of A satisfy the equation of the transformed graph.

4 For each of the graphs below,

    (i) state two transformations that will transform $y = \cos x$ (shown in grey on each diagram) on to the given graph

    (ii) hence write down the equation of the graph

(a)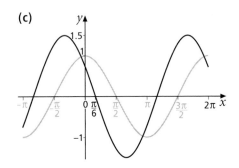

(b)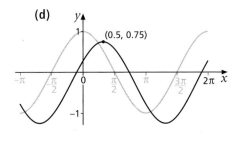

(c)

(d)

5 For $-2\pi \leq x \leq 2\pi$ sketch the graphs of

    (a) $y = \sin x$ **(b)** $y = |\sin x|$ **(c)** $y = \sin|x|$

    Check using a graph plotter.

6 (a) Sketch the graph of $y = \sec x$ for $-\pi \leq x \leq \pi$.

(b) On the same diagram, sketch the graph of $y = \sec x$ after a stretch in the y-direction, scale factor $\frac{1}{2}$, followed by a stretch in the x-direction, scale factor $\frac{1}{3}$.

(c) What is the equation of the resulting graph? Check with a graph plotter.

7 The graph of $y = \cos 3x$ is shown $(-\pi \leq x \leq \pi)$.

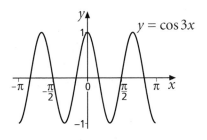

(a) Identify the coordinates of each maximum and minimum of $y = \cos 3x$ in the interval $-\pi \leq x \leq \pi$.

(b) Copy the graph and on your copy draw the graph of $y = |\cos 3x|$.

(c) Use your answers to (a) to solve the equation $|\cos 3x| = 1$ $(-\pi \leq x \leq \pi)$.

8 (a) Find the equation of the resulting graph when $y = \arcsin x$ is first stretched, scale factor 2, in the y-direction and then translated by $\begin{bmatrix} 1 \\ 0 \end{bmatrix}$.

(b) Sketch the resulting graph.
Check using a graph plotter.

9 $f(x)$ is defined by $f(x) = \sin\frac{1}{2}x$ for $x \in \mathbb{R}, -\pi \leq x \leq \pi$;
$g(x)$ is defined by $g(x) = |x|$ for $x \in \mathbb{R}$.

(a) State the range of $f(x)$.

(b) State the domain and range of $f^{-1}(x)$.

(c) Which of the following functions is $f^{-1}(x)$?

$$\boxed{f^{-1}(x) = \arcsin 2x} \quad \boxed{f^{-1}(x) = \arcsin\tfrac{1}{2}x} \quad \boxed{f^{-1}(x) = 2\arcsin x} \quad \boxed{f^{-1}(x) = \tfrac{1}{2}\arcsin x}$$

(d) Write down an expression for $fg(x)$.

(e) Sketch $y = fg(x)$.

*10 Use a graph plotter for this question.

(a) Draw the graph of $y = \tan x$ and $y = \cot x$ on the same axes.

(b) The two graphs appear to be the same shape.
Perform two transformations on the graph of $y = \cot x$ to superimpose the graph on to that of $y = \tan x$ to confirm this observation.

(c) Write down the equation of the graph of $y = \cot x$ after the two transformations in part (b).

(d) Prove algebraically that your equation in (c) is identical to $y = \tan x$.

Key points

- The inverse function of $\sin x$ is denoted by $y = \arcsin x$, $\sin^{-1} x$ or $\operatorname{asin} x$, and has domain $-1 \leq x \leq 1$ and range $-\dfrac{\pi}{2} \leq y \leq \dfrac{\pi}{2}$.

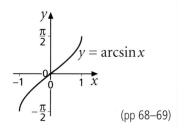

(pp 68–69)

- The inverse function of $\cos x$ is denoted by $y = \arccos x$, $\cos^{-1} x$ or $\operatorname{acos} x$, and has domain $-1 \leq x \leq 1$ and range $0 \leq y \leq \pi$.

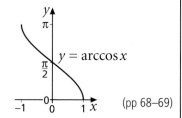

(pp 68–69)

- The inverse function of $\tan x$ is denoted by $y = \arctan x$, $\tan^{-1} x$ or $\operatorname{atan} x$, and has domain $x \in \mathbb{R}$ and range $-\dfrac{\pi}{2} < y < \dfrac{\pi}{2}$.

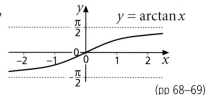

(pp 68–69)

- $\sec x$ is defined as $\dfrac{1}{\cos x}$.

  $y = \sec x$ has domain $x \in \mathbb{R}$, $x \neq \pm\dfrac{\pi}{2}, \pm\dfrac{3\pi}{2}, \pm\dfrac{5\pi}{2}, \ldots$

  and range $y \leq -1$ and $y \geq 1$;

  its period is 2π.

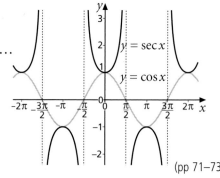

(pp 71–73)

- $\operatorname{cosec} x$ is defined as $\dfrac{1}{\sin x}$.

  $y = \operatorname{cosec} x$ has domain $x \in \mathbb{R}$, $x \neq 0, \pm\pi, \pm 2\pi, \pm 3\pi, \ldots$

  and range $y \leq -1$ and $y \geq 1$;

  its period is 2π.

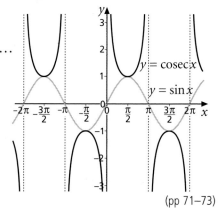

(pp 71–73)

- $\cot x$ is defined as $\dfrac{1}{\tan x}$.

  $y = \cot x$ has domain $x \in \mathbb{R}, x \neq 0, \pm\pi, \pm2\pi, \pm3\pi, \ldots$
  and range $y \in \mathbb{R}$; its period is π.
  ($\tan x$ is shown in grey on the graph.)

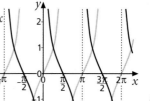

  (pp 71–73)

- $\sec^2 x = 1 + \tan^2 x$ (p 73)

- $\csc^2 x = 1 + \cot^2 x$ (p 73)

Test yourself (answers p 181)

1 Write down, in radians, the exact value of

   (a) $\arcsin \frac{1}{2}$ **(b)** $\arctan\left(\dfrac{1}{\sqrt{3}}\right)$ **(c)** $\arccos\left(-\dfrac{\sqrt{3}}{2}\right)$ **(d)** $\arccos\left(-\frac{1}{2}\right)$

2 Give exact values of

   (a) $\csc 30°$ **(b)** $\sec 45°$ **(c)** $\cot 60°$ **(d)** $\sec (-60°)$

3 Simplify these.

   (a) $\cot x \sec x$ **(b)** $\dfrac{1}{(1 - \csc x)(1 + \csc x)}$ **(c)** $\dfrac{\sec^2 x}{(\sec x - 1)(1 + \sec x)}$

4 Solve $\csc^2 x + \cot x - 1 = 0$ giving answers as multiples of π in the range 0 to 2π.

5 Solve the equation $3 \cot^2 x + 8 \csc x + 1 = 0$,
   giving all values of x to the nearest degree in the interval $0° \leq x \leq 360°$.

6 State the two transformations that will map $y = \tan x$ on to

   (a) $y = 2\tan(-x)$ **(b)** $y = \tan(2x) + 1$ **(c)** $y = -\tan\left(\frac{1}{2}x\right)$ **(d)** $y = 2 - \tan x$

7 What is the equation of the image of $y = \sin x$ after each of these?

   (a) A stretch in the x-direction, factor 4, followed by a translation by $\begin{bmatrix} 0 \\ 2 \end{bmatrix}$

   (b) A reflection in the y-axis, followed by a stretch in the y-direction, factor 2

   (c) A translation by $\begin{bmatrix} -1 \\ 0 \end{bmatrix}$ followed by a reflection in the x-axis

8 For each of the graphs below,

     (i) state the two transformations that will transform $y = \sin x$
         on to the given graph ($y = \sin x$ is shown in grey on each diagram)

     (ii) hence write down the equation of the graph

(a)

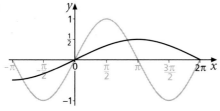

(b)

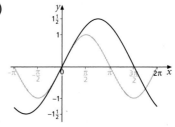

6 Trigonometric formulae

In this chapter you will learn how to
- use the formulae for $\sin(A + B)$, $\cos(A + B)$, etc.
- use the formulae for $\sin 2A$, $\cos 2A$ and $\tan 2A$
- use equivalent expressions for $a\cos\theta + b\sin\theta$
- solve equations of the form $a\cos\theta + b\sin\theta = c$

A Addition formulae (answers p 182)

The diagram shows a rectangle with diagonal 1 unit long, touching two parallel lines.

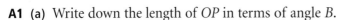

A1 (a) Write down the length of OP in terms of angle B.

(b) Use your answer to (a) to write down an expression for the length of RP.

(c) Show that angle QPT = angle A, and use this to find an expression for the length of PT.

(d) Hence write down an expression for $RP + PT$, the distance between the parallel lines.

(e) Find QH in terms of angle $(A + B)$, and thus show that
$\sin(A + B) = \sin A \cos B + \cos A \sin B$.

A2 By using the fact that $OH = OR - HR$, show that $\cos(A + B) = \cos A \cos B - \sin A \sin B$.

The formulae you derived in questions A1 and A2 are called the **addition formulae**.

A3 By replacing B by $-B$ in the two addition formulae, and simplifying, establish formulae for $\sin(A - B)$ and $\cos(A - B)$.

> **K**
> $\sin(A + B) = \sin A \cos B + \cos A \sin B$
> $\sin(A - B) = \sin A \cos B - \cos A \sin B$
> $\cos(A + B) = \cos A \cos B - \sin A \sin B$
> $\cos(A - B) = \cos A \cos B + \sin A \sin B$

A4 Put $B = A$ in the formula for $\sin(A + B)$. What result do you get for $\sin 2A$?

A5 By putting $B = A$ in the formula for $\cos(A + B)$,

(a) show that $\cos 2A = 2\cos^2 A - 1$

(b) show also that $\cos 2A = 1 - 2\sin^2 A$

> **K**
> $\sin 2A = 2\sin A \cos A$
> $\cos 2A = \cos^2 A - \sin^2 A = 2\cos^2 A - 1 = 1 - 2\sin^2 A$

These are sometimes referred to as double angle formulae.

Example 1

Use an addition formula to show that $\cos(x + 90)° = -\sin x°$.
Explain your result graphically.

Solution

In the addition formula for $\cos(A + B)$,
substitute $x°$ for A and $90°$ for B.

$$\cos(x + 90)° = \cos x° \cos 90° - \sin x° \sin 90°$$
$$= \cos x° \times 0 - \sin x° \times 1$$
$$= -\sin x°$$

The graph of $\cos(x + 90)°$ is
the graph of $\cos x°$ translated by $\begin{bmatrix} -90° \\ 0 \end{bmatrix}$.

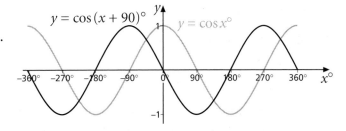

The graph of $-\sin x°$ is the graph
of $\sin x°$ reflected in the x-axis.

The two graphs are clearly identical,
illustrating that $\cos(x + 90)° = -\sin x°$.

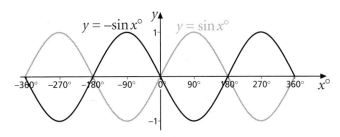

Example 2

By writing $\cos 15°$ as $\cos(60° - 45°)$, show that $\cos 15° = \dfrac{1 + \sqrt{3}}{2\sqrt{2}}$.

Solution

$$\cos 15° = \cos(60° - 45°)$$
$$= \cos 60° \cos 45° + \sin 60° \sin 45°$$
$$= \tfrac{1}{2} \times \frac{1}{\sqrt{2}} + \frac{\sqrt{3}}{2} \times \frac{1}{\sqrt{2}}$$
$$= \frac{1 + \sqrt{3}}{2\sqrt{2}}$$

Example 3

Given that $\sin A = \tfrac{1}{4}$, find   (a) $\cos 2A$ (b) $\sin 2A$

Solution

(a) $\cos 2A = 1 - 2\sin^2 A = 1 - 2 \times \tfrac{1}{16} = 1 - \tfrac{1}{8} = \tfrac{7}{8}$

(b) Since $\sin^2 2A + \cos^2 2A = 1$, $\sin 2A = \pm\sqrt{1 - \cos^2 2A}$

so $\sin 2A = \pm\sqrt{1 - \dfrac{49}{64}} = \pm\sqrt{\dfrac{15}{64}} = \pm\dfrac{\sqrt{15}}{8}$

D **A6** In example 3, there is only one value for $\cos 2A$, but two values for $\sin 2A$. Explain why this is so.

There are addition formulae for the tangent function as well as sine and cosine.

A7 **(a)** Rewrite $\tan(A + B)$ in terms of $\sin(A + B)$ and $\cos(A + B)$.

(b) Use the addition formulae to expand $\sin(A + B)$ and $\cos(A + B)$ in the expression you wrote in part (a).

(c) Divide top and bottom of the fraction by $\cos A \cos B$, to obtain an expression solely in terms of $\tan A$ and $\tan B$.

A8 **(a)** Replace B by $-B$ to obtain an expression for $\tan(A - B)$ in terms of $\tan A$ and $\tan B$.

(b) In the expression for $\tan(A + B)$, put $B = A$ and thus obtain an expression for $\tan 2A$ in terms of $\tan A$.

K

$$\tan(A + B) = \frac{\tan A + \tan B}{1 - \tan A \tan B}$$

$$\tan(A - B) = \frac{\tan A - \tan B}{1 + \tan A \tan B}$$

$$\tan 2A = \frac{2 \tan A}{1 - \tan^2 A} \qquad \text{(This is another double angle formula.)}$$

Example 4

Given that $\tan 2A = 2$, find in surd form the two possible values of $\tan A$.

Solution

$$\tan 2A = \frac{2 \tan A}{1 - \tan^2 A} = 2$$

$$\Rightarrow \qquad 2 \tan A = 2(1 - \tan^2 A)$$

$$\Rightarrow \qquad \tan A = 1 - \tan^2 A$$

$$\Rightarrow \qquad \tan^2 A + \tan A - 1 = 0$$

Use the quadratic formula.
$$\tan A = \frac{-1 \pm \sqrt{1^2 - 4 \times 1 \times -1}}{2 \times 1}$$

$$\Rightarrow \qquad \tan A = \frac{-1 \pm \sqrt{5}}{2}$$

$$\Rightarrow \qquad \tan A = -\tfrac{1}{2} + \frac{\sqrt{5}}{2} \quad \text{or} \quad -\tfrac{1}{2} - \frac{\sqrt{5}}{2}$$

Example 5

Find a formula for $\tan 3A$ in terms of $\tan A$.

Solution

Use the addition formula.
$$\tan 3A = \tan(A + 2A) = \frac{\tan A + \tan 2A}{1 - \tan A \tan 2A}$$

Use the double angle formula.
$$= \frac{t + \dfrac{2t}{1 - t^2}}{1 - t\dfrac{2t}{1 - t^2}} \quad \text{where } t = \tan A$$

Multiply top and bottom by $(1 - t^2)$.
$$= \frac{t(1 - t^2) + 2t}{(1 - t^2) - t \times 2t}$$

$$= \frac{3t - t^3}{1 - 3t^2} = \frac{3\tan A - \tan^3 A}{1 - 3\tan^2 A}$$

Example 6

Solve $3\sin 2x = \cos x$, giving answers between 0 and 2π in radians to two decimal places.

Solution

$$3\sin 2x = \cos x$$
$$\Rightarrow \quad 3 \times 2 \sin x \cos x = \cos x$$
$$\Rightarrow \quad 6 \sin x \cos x - \cos x = 0$$
$$\Rightarrow \quad \cos x(6 \sin x - 1) = 0$$

So $\cos x = 0$, or $6\sin x - 1 = 0$ giving $\sin x = \frac{1}{6}$

$\cos x = 0$ gives $x = \dfrac{\pi}{2}$ or $\dfrac{3\pi}{2}$, i.e. $x = 1.57$ or 4.71 (to 2 d.p.)

$\sin x = \frac{1}{6}$ gives $x = \arcsin\frac{1}{6} = 0.167\ldots$, or $\pi - 0.167\ldots$ which is $2.974\ldots$

So $x = 0.17, 1.57, 2.97$ or 4.71 (to 2 d.p.)

Example 7

Prove that $\dfrac{1 - \cos 2A}{\sin 2A} = \tan A \quad (A \neq n\pi, n \in \mathbb{Z})$

Solution

$$\frac{1 - \cos 2A}{\sin 2A} = \frac{1 - (1 - 2\sin^2 A)}{2\sin A \cos A} = \frac{2\sin^2 A}{2\sin A \cos A} = \frac{\sin A}{\cos A} = \tan A$$

$n \in \mathbb{Z}$ *means 'n is a member of the integers'* $(\ldots, -2, -1, 0, 1, 2, \ldots)$
So $(A \neq n\pi, n \in \mathbb{Z})$ means that A cannot be $\ldots, -2\pi, -\pi, 0, \pi, 2\pi, \ldots$

*These are the values of A where $\sin 2A = 0$ and where $\tan A$ is not defined.
You are not expected to prove that the equality does not hold for these values;
the question is simply telling you that it does not hold.*

Example 8

Show that $\cos x - \sin x - 1 = -2\sin\frac{x}{2}\left(\sin\frac{x}{2} + \cos\frac{x}{2}\right)$.

Solution

$\cos 2A = 1 - 2\sin^2 A$. Replacing A by $\frac{x}{2}$, we have $\cos x = 1 - 2\sin^2\frac{x}{2}$.

Similarly $\sin 2A = 2\sin A\cos A$. Replacing A by $\frac{x}{2}$, we have $\sin x = 2\sin\frac{x}{2}\cos\frac{x}{2}$.

Hence $\cos x - \sin x - 1 = \left(1 - 2\sin^2\frac{x}{2}\right) - 2\sin\frac{x}{2}\cos\frac{x}{2} - 1$

$$= -2\sin^2\frac{x}{2} - 2\sin\frac{x}{2}\cos\frac{x}{2}$$

$$= -2\sin\frac{x}{2}\left(\sin\frac{x}{2} + \cos\frac{x}{2}\right)$$

Exercise A (answers p 182)

1 Use an addition formula to simplify $\sin(x + 180)°$.
Explain your result graphically.

2 (a) By writing $\sin 75°$ as $\sin(45° + 30°)$, show that $\sin 75° = \dfrac{\sqrt{3}+1}{2\sqrt{2}}$.

(b) Use a similar method to express $\sin 15°$ using surds.

3 Given that $\cos A = \frac{4}{5}$, where A is acute, find the exact value of

(a) $\sin A$ **(b)** $\sin 2A$ **(c)** $\cos 2A$ **(d)** $\operatorname{cosec} 2A$

4 (a) If $\tan A = \frac{1}{2}$ and $\tan B = \frac{1}{3}$, find $\tan(A + B)$.

(b) If $\tan 2C = \frac{3}{4}$, find the two possible values of $\tan C$.

5 (a) Show that $\cos(A + B) + \cos(A - B) = 2\cos A\cos B$.

(b) Simplify $\cos(A - B) - \cos(A + B)$.

6 Simplify

(a) $\cos 2A\cos A - \sin 2A\sin A$ **(b)** $\cos(A + B)\cos A + \sin(A + B)\sin A$

(c) $2\sin 3C\cos 3C$ **(d)** $\sin 3D\cos 2D + \cos 3D\sin 2D$

7 Show that

(a) $(\cos A + \sin A)(\cos B + \sin B) = \cos(A - B) + \sin(A + B)$

(b) $(\cos A + \sin A)^2 = 1 + \sin 2A$

(c) $\sin 3A = 3\sin A - 4\sin^3 A$

(d) $(\sin A + \cos B)^2 + (\cos A - \sin B)^2 = 2(1 + \sin(A - B))$

8 Show that $\cot 2x + \operatorname{cosec} 2x = \cot x$.

9 Show that $\dfrac{\cos\theta}{1-\sqrt{2}\sin\theta} - \dfrac{\cos\theta}{1+\sqrt{2}\sin\theta} = \sqrt{2}\tan 2\theta \ \left(\theta \neq n\pi \pm \dfrac{\pi}{4}, n \in \mathbb{Z}\right)$.

10 Prove that $\dfrac{\tan a}{\sec a - 1} = \cot \frac{1}{2}a \ (a \neq n\pi, n \in \mathbb{Z})$.

11 If $t = \tan\frac{1}{2}\theta$, show that

(a) $\dfrac{2t}{1+t^2} = \sin\theta$ (b) $\dfrac{1-t^2}{1+t^2} = \cos\theta$

12 Solve these equations for $0° \leq x° \leq 360°$.

(a) $\sin 2x° = \cos x°$ (b) $\sin x° + \cos 2x° = 0$ (c) $\cos 2x° = 7\cos x° + 3$

(d) $\cos 2x° = 1 + \sin x°$ (e) $\sin 2x° = \tan x°$

13 Show that

(a) $\sin(\theta + \phi) + \sin(\theta - \phi) = 2\sin\theta\cos\phi$

(b) $\sin(\theta + \phi) - \sin(\theta - \phi) = 2\cos\theta\sin\phi$

14 (a) Write down formulae for $\sin(A + B)$ and $\sin(A - B)$.

Using $X = A + B$ and $Y = A - B$, prove that $\sin X + \sin Y = 2\sin\dfrac{X+Y}{2}\cos\dfrac{X-Y}{2}$.

(b) Hence or otherwise, solve, for $0° \leq \theta \leq 360°$, $\sin 4\theta + \sin 2\theta = 0$.   Edexcel

B Equivalent expressions (answers p 185)

You already know how to solve several types of trigonometric equations.

B1 Solve each of these equations to the nearest degree. Give all solutions between 0° and 360°.
Check that each of your answers is reasonable by using a graph plotter.

(a) $2\sin(x + 30)° = 1$ (b) $3\cos(x - 50)° = 2$ (c) $3\cos x° + 4\sin x° = 0$

B2 Can you solve $3\sin x° + 4\cos x° = 1$ using any of the methods you used in B1? Explain your answer.

In order to solve $3\sin x° + 4\cos x° = 1$, it will be useful to look at the trigonometric function $y = 3\sin x° + 4\cos x°$.

B3 (a) On a graph plotter, draw the graph of $y = 3\sin x° + 4\cos x°$, for values of $x°$ roughly between $-360°$ and $360°$.

(b) The graph of $y = 3\sin x° + 4\cos x°$ appears to be a sine curve.

(i) What is the period of $y = 3\sin x° + 4\cos x°$?

(ii) What is the amplitude of $y = 3\sin x° + 4\cos x°$?

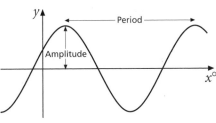

B4 Using a graph plotter, find the period and amplitude of each of these.

(a) $y = 5 \sin x° + 12 \cos x°$ (b) $y = 7 \sin x° + 24 \cos x°$ (c) $y = \sin x° + \cos x°$

B5 Look at your results in questions B3 and B4.

(a) What do you think the period of $y = 2 \sin x° + 3 \cos x°$ would be?

(b) What do you think the amplitude of $y = 2 \sin x° + 3 \cos x°$ would be?

(c) Check your conjectures by plotting the graph of $y = 2 \sin x° + 3 \cos x°$.

(d) What do you think the period and amplitude of $y = a \sin x° + b \cos x°$ would be?

B6 (a) Draw again the graph of $y = \sin x° + \cos x°$ on your plotter.
Draw also the graph of $y = \sin x°$.

(b) It looks as though you can transform the graph of $y = \sin x°$ on to
that of $y = \sin x° + \cos x°$ using two transformations:

an enlargement in the y-direction scale factor $\sqrt{2}$, followed by a translation of $\begin{bmatrix} -45° \\ 0 \end{bmatrix}$.

What does the equation of $y = \sin x°$ become after these two transformations?

(c) Use the fact that $\sin (x + \alpha)° = \sin x° \cos \alpha° + \cos x° \sin \alpha°$ to show that the
equation of the graph you obtained in part (b) is identical to $y = \sin x° + \cos x°$.

K
$$r \sin (x + \alpha) = r(\sin x \cos \alpha + \cos x \sin \alpha)$$
$$= (r \cos \alpha) \sin x + (r \sin \alpha) \cos x$$

If we have an expression such as $2 \sin x + 3 \cos x$, we can change it into the form
$r \sin (x + \alpha)$ provided we can find values of r and α, such that

$$r \cos \alpha = 2 \text{ and } r \sin \alpha = 3$$

An example will show how this can be useful.

Example 9

Find the maximum value of the expression $2 \sin x° + 3 \cos x°$,
and the value of $x°$ ($0° \le x° \le 360°$) at which the maximum occurs.

Solution

First express $2 \sin x° + 3 \cos x°$ *in the form* $r \sin (x + \alpha)°$.

If $2 \sin x° + 3 \cos x° = r \sin x° \cos \alpha° + r \cos x° \sin \alpha°$ then,
equating coefficients, $r \cos \alpha° = 2$ and $r \sin \alpha° = 3$.

Eliminate $\alpha°$ *from these two equations by squaring and adding.*

$$r^2 \cos^2 \alpha° + r^2 \sin^2 \alpha° = 4 + 9 = 13$$

$\Rightarrow \qquad r^2 (\cos^2 \alpha° + \sin^2 \alpha°) = 13$

$\Rightarrow \qquad\qquad\qquad r^2 = 13 \qquad$ *since* $\cos^2 \alpha° + \sin^2 \alpha° = 1$.

$\Rightarrow \qquad\qquad\qquad r = \sqrt{13} \qquad$ *Take the positive value.*

You could use this value of r to find the value of $\alpha°$ *using* $r \cos \alpha° = 2$ *or* $r \sin \alpha° = 3$.
But you can easily find $\alpha°$ *by eliminating r directly from* $r \cos \alpha° = 2$ *and* $r \sin \alpha° = 3$.

$r \cos \alpha° = 2$ and $r \sin \alpha° = 3$

$\Rightarrow \dfrac{r \sin \alpha°}{r \cos \alpha°} = \dfrac{3}{2} \Rightarrow \tan \alpha° = 1.5 \Rightarrow \alpha° = 56°$ (to the nearest degree)

Hence $2 \sin x° + 3 \cos x° = \sqrt{13} \sin (x + 56)°$

This expression has a maximum value of $\sqrt{13}$, when $\sin (x + 56)° = 1$.
$\sin (x + 56)° = 1$ when $(x + 56)° = 90°$

Hence $x° = 34°$ *There is only one solution for x between 0° and 360°.*

B7 In the example above, $r^2 = 13$. Hence r could be $-\sqrt{13}$.
In this case, $\cos \alpha°$ and $\sin \alpha°$ will both be negative, so $180° \leq \alpha° \leq 270°$.
Using $r = -\sqrt{13}$ and an appropriate value of $\alpha°$, check that this
leads to the same solution as above.

An expression such as $4 \sin x° - 3 \cos x°$ can be put in the form $r \sin (x - \alpha)°$.

B8 (a) Expand the expression $r \sin (x - \alpha)°$ using the addition formula.

(b) Given that $4 \sin x° - 3 \cos x° = r \sin (x - \alpha)°$, by equating coefficients
and squaring, find the value of r, where $r > 0$.

(c) Similarly, show that $\tan \alpha° = 0.75$ and hence find $\alpha°$ to the nearest degree.

(d) Hence express $4 \sin x° - 3 \cos x°$ in the form $r \sin (x - \alpha)°$.

(e) (i) What is the minimum value of $4 \sin x° - 3 \cos x°$?

(ii) At what value of $x°$, between 0° and 360°, does the minimum occur?

(f) Check that your answer is correct by plotting $y = 4 \sin x° - 3 \cos x°$
and your answer to (d) using a graph plotter.

You can also use the expressions $r \cos (x + \alpha)$ and $r \cos (x - \alpha)$ in the same way.

Example 10

(a) Express $4 \cos x + 3 \sin x$ in the form $R \cos (x - \alpha)$, where $R > 0$ and $0 < \alpha < \dfrac{\pi}{2}$.

(b) Hence solve $4 \cos x + 3 \sin x = 2$ for $0 < x < \pi$, giving answers to two decimal places.

Solution

(a) $R \cos (x - \alpha) = R \cos x \cos \alpha + R \sin x \sin \alpha$

If $4 \cos x + 3 \sin x = R \cos (x - \alpha)$, we require $R \cos \alpha = 4$ and $R \sin \alpha = 3$.

Hence $\tan \alpha = \dfrac{3}{4}$, $\alpha = \tan^{-1} 0.75 = 0.64$ radians (to 2 d.p.) and $R^2 = 4^2 + 3^2 = 25$, so $R = 5$.

So $4 \cos x + 3 \sin x = 5 \cos (x - 0.64)$.

(b) $4 \cos x + 3 \sin x = 2 \qquad \Rightarrow \quad 5 \cos (x - 0.64) = 2$

$\Rightarrow \qquad \cos (x - 0.64) = 0.4$

One solution is given by $\qquad\qquad x - 0.64 = \cos^{-1} 0.4 = 1.159\ldots$

$\Rightarrow \qquad\qquad\qquad x = 1.799\ldots = 1.80$ (to 2 d.p.)

This is the only solution between 0 and π.

Example 11

(a) Express $\cos\theta - \sin\theta$ in the form $r\cos(\theta + \alpha)$.

(b) Hence solve the equation $\cos\theta - \sin\theta = 0.5$ for $0 \le \theta \le \dfrac{\pi}{2}$.

Solution

(a) Let $\cos\theta - \sin\theta = r\cos(\theta + \alpha)$

$$= r(\cos\theta\cos\alpha - \sin\theta\sin\alpha)$$

$$= (r\cos\alpha)\cos\theta - (r\sin\alpha)\sin\theta$$

Then, equating coefficients, we have $r\cos\alpha = 1$ and $r\sin\alpha = 1$.

Hence $r^2 = 1^2 + 1^2 = 2$, so $r = \sqrt{2}$.

$$\frac{r\sin\alpha}{r\cos\alpha} = \frac{1}{1} \Rightarrow \tan\alpha = 1, \text{ so } \alpha = \frac{\pi}{4}$$

Hence $\cos\theta - \sin\theta = \sqrt{2}\cos\left(\theta + \dfrac{\pi}{4}\right)$

(b) Thus $\cos\theta - \sin\theta = 0.5 \Rightarrow \sqrt{2}\cos\left(\theta + \dfrac{\pi}{4}\right) = 0.5 \Rightarrow \cos\left(\theta + \dfrac{\pi}{4}\right) = \dfrac{0.5}{\sqrt{2}} = 0.3536\ldots$

$\Rightarrow \left(\theta + \dfrac{\pi}{4}\right) = 1.209\ldots$ *Remember you are working in radians.*

$\Rightarrow \qquad \theta = 1.209\ldots - \dfrac{\pi}{4} = 0.42$ (to 2 d.p.)   *This is the only solution between 0 and $\dfrac{\pi}{2}$.*

Example 12

Solve the equation $9\sin\theta° - 6\cos\theta° = 7$ for $0° \le \theta° \le 360°$.

Solution

$9\sin\theta° - 6\cos\theta°$ could be expressed in the form $r\sin(\theta - \alpha)°$ or $-r\cos(\theta + \alpha)°$; either would be suitable. Here $r\sin(\theta - \alpha)°$ is used.

Let $9\sin\theta° - 6\cos\theta° = r\sin(\theta - \alpha)° = r\sin\theta°\cos\alpha° - r\cos\theta°\sin\alpha°$.

Hence $r\cos\alpha° = 9$ and $r\sin\alpha° = 6$

$\Rightarrow r^2 = 9^2 + 6^2 = 117$, so $r = 10.82\ldots$

Also $\dfrac{r\sin\alpha°}{r\cos\alpha°} = \dfrac{6}{9} \Rightarrow \tan\alpha° = 0.666\ldots$, from which $\alpha° = 33.69\ldots°$

So $9\sin\theta° - 6\cos\theta° = 10.82\ldots \times \sin(\theta - 33.69)° = 7$

$\Rightarrow \sin(\theta - 33.69)° = 7 \div 10.82\ldots = 0.6472\ldots$

$\theta°$ is between $0°$ and $360°$, so $(\theta - 33.69)°$ is between $-33.69°$ and $326.31°$.

$\sin(\theta - 33.69)° = 0.6472\ldots$ gives $(\theta - 33.69)° = 40.33°$ (to the nearest $0.01°$)

$$\text{and } (\theta - 33.69)° = 180° - 40.33° = 139.67°$$

We also have $(\theta - 33.69)° = 40.33° + 360°$ etc., but these give values of $\theta°$ outside the required range.

The two values above are within the range $-33.69°$ to $326.31°$, and they lead to

$$(\theta - 33.69)° = 40.33° \qquad \Rightarrow \qquad \theta = 74.0°$$

$$\text{and } (\theta - 33.69)° = 139.67° \qquad \Rightarrow \qquad \theta = 173.4° \text{ (answers correct to } 0.1°)$$

Exercise B (answers p 186)

1 The graph shown may be regarded either as a sine graph or a cosine graph, translated left or right.
Express the equation of the graph in the following forms, where $r > 0$ and $0° < \alpha° < 360°$.

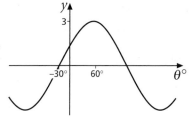

(a) $r\sin(\theta + \alpha)°$ (b) $r\cos(\theta - \alpha)°$

(c) $r\sin(\theta - \alpha)°$ (d) $r\cos(\theta + \alpha)°$

2 (a) Express $5\sin\theta° + 2\cos\theta°$ in the form $r\sin(\theta + \alpha)°$, where $r > 0$ and $0° < \alpha° < 90°$.

   (b) Hence find the maximum value of $5\sin\theta° + 2\cos\theta°$, and the value of $\theta°$ $(0° < \theta° < 90°)$ at which it occurs.

3 (a) Express $\sin\theta° + 2\cos\theta°$ in the form $r\sin(\theta + \alpha)°$, where $r > 0$ and $0° < \alpha° < 90°$.

   (b) Express $\sin\theta° + 2\cos\theta°$ in the form $r\cos(\theta - \alpha)°$, where $r > 0$ and $0° < \alpha° < 90°$.

   (c) Show that your answers in parts (a) and (b) are equivalent.

4 (a) Express $2\sin\theta° - 3\cos\theta°$ in the form $r\sin(\theta - \alpha)°$, where $r > 0$ and $0° < \alpha° < 90°$.

   (b) By writing $2\sin\theta° - 3\cos\theta°$ as $-(3\cos\theta° - 2\sin\theta°)$, write $2\sin\theta° - 3\cos\theta°$ in the form $-r\cos(\theta + \alpha)°$, where $r > 0$ and $0° < \alpha° < 90°$.

   (c) Using your answer to (a) or to (b), solve the equation $2\sin\theta° - 3\cos\theta° = 1$ for $0° < \theta° < 90°$.

5 Solve each of these equations for $0° < x° < 360°$.

   (a) $7\sin x° + 10\cos x° = 8$ (b) $3\sin x° - 4\cos x° = 2$ (c) $9\cos x° - 5\sin x° = 4$

6 Prove that if $\cos(x + 45)° = 2\cos(x - 45)°$, then $\tan x° = -\frac{1}{3}$.

7 (a) Express $\sqrt{3}\sin\theta - \cos\theta$ in the form in the form $r\sin(\theta - \alpha)$ $\left(r > 0 \text{ and } 0 < \alpha < \frac{\pi}{2}\right)$.

   (b) Hence solve $\sqrt{3}\sin\theta - \cos\theta = 1$, giving exact solutions between 0 and 2π.

8 (a) Show that $\tan(\theta - 45)° = \dfrac{\tan\theta° - 1}{1 + \tan\theta°}$.

   (b) Hence find the exact value of $\tan 15°$, giving your answer in the form $p + q\sqrt{3}$.

9 (a) Given that $\cos A = \frac{5}{13}$ $\left(0 < A < \frac{\pi}{2}\right)$, find the exact value of $\sin A$.

   (b) Given that $\sin B = \frac{3}{5}$ $\left(0 < B < \frac{\pi}{2}\right)$, find the exact value of $\cos(A + B)$.

10 (a) (i) Express $(12\cos\theta - 5\sin\theta)$ in the form $R\cos(\theta + \alpha)$, where $R > 0$ and $0 < \alpha < 90°$.

     (ii) Hence solve the equation $12\cos\theta - 5\sin\theta = 4$, for $0 < \theta < 90°$, giving your answer to one decimal place.

   (b) Solve $8\cot\theta - 3\tan\theta = 2$, for $0 < \theta < 90°$, giving your answer to one decimal place.

Edexcel

- $\sin(A + B) = \sin A \cos B + \cos A \sin B$
  $\sin(A - B) = \sin A \cos B - \cos A \sin B$
  $\cos(A + B) = \cos A \cos B - \sin A \sin B$
  $\cos(A - B) = \cos A \cos B + \sin A \sin B$ (p 84)

- $\tan(A + B) = \dfrac{\tan A + \tan B}{1 - \tan A \tan B}$ $\qquad \tan(A - B) = \dfrac{\tan A - \tan B}{1 + \tan A \tan B}$ (p 86)

- $\sin 2A = 2 \sin A \cos A$
  $\cos 2A = \cos^2 A - \sin^2 A = 2 \cos^2 A - 1 = 1 - 2 \sin^2 A$ (p 84)

- $\tan 2A = \dfrac{2 \tan A}{1 - \tan^2 A}$ (p 86)

- $a \sin x + b \cos x$ can be written in the form
  $r \sin(x + \alpha)$, where $a = r \cos \alpha$ and $b = r \sin \alpha$
  $r \cos(x - \alpha)$, where $a = r \sin \alpha$ and $b = r \cos \alpha$ (pp 90–91)

- $a \sin x - b \cos x$ can be written in the form
  $r \sin(x - \alpha)$, where $a = r \cos \alpha$ and $b = r \sin \alpha$
  $r \cos(x + \alpha)$, where $a = -r \sin \alpha$ and $b = -r \cos \alpha$ (p 91)

Test yourself (answers p 186)

1 Use an addition formula to simplify $\sin(x + 90)°$, and explain your result with the aid of a sketch graph.

2 Express the equation of the graph shown in the following forms, where $r > 0$ and $0° < \alpha° < 360°$.

(a) $r \sin(\theta + \alpha)°$ (b) $r \cos(\theta - \alpha)°$

(c) $r \sin(\theta - \alpha)°$ (d) $r \cos(\theta + \alpha)°$

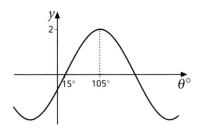

3 (a) By writing $\cos 105°$ as $\cos(60 + 45)°$, show that $\cos 105° = \dfrac{1 - \sqrt{3}}{2\sqrt{2}}$.

   (b) Use a similar method to express $\sin 105°$ using surds.

4 Given that $\sin x = \frac{3}{5}$, where $0 < x < \dfrac{\pi}{2}$, find exactly

(a) $\cos x$ (b) $\cos 2x$ (c) $\sin 2x$ (d) $\tan 2x$

5 Given that $\sin\theta = \frac{12}{13}$, where $\frac{\pi}{2} < \theta < \pi$, find the exact value of

(a) $\cos\theta$ (b) $\cos 2\theta$ (c) $\operatorname{cosec} 2\theta$ (d) $\cot 2\theta$

6 Given that $\tan 2\theta = 3$, find the two possible values of $\tan\theta$ in surd form.

7 Solve $\cos^2\theta° = \sin 2\theta°$ for $0° < \theta° < 360°$.

8 Solve these equations, giving your solutions in radians between 0 and 2π in terms of π or (where that is not possible) to two decimal places.

(a) $2\cos x = \sin 2x$ (b) $\cos 2x = 1 + \sin 2x$ (c) $\cos 2x = 3\cos x + 2$

9 Prove that $\cot\dfrac{\theta}{2} - \tan\dfrac{\theta}{2} = 2\cot\theta$.

10 Show that $\tan(A+B) + \tan(A-B) = 2\cot A\,\dfrac{\sec^2 B}{\cot^2 A - \tan^2 B}$.

11 (a) Write down the expansion of the expression $r\sin(x+\alpha)°$.

(b) Hence express $3\sin x° + 5\cos x°$ in the form $r\sin(x+\alpha)°$, where $r > 0$ and $0° < \alpha° < 90°$.

(c) Thus find the maximum value of $3\sin x° + 5\cos x°$ and the value of $x°$ $(0° < x° < 180°)$ at which it occurs.

12 (a) Express $3\sin x° + 4\cos x°$ in the form $r\sin(x+\alpha)°$, where $r > 0$ and $0° < \alpha° < 90°$.

(b) Hence find the maximum and minimum values of the expression
$$\frac{4}{3\sin x° + 4\cos x° + 11}$$
and the values of $x°$ at which they occur $(0° < x° < 360°)$.

13 (a) (i) Express $5\sin\theta° - 4\cos\theta°$ in the form $r\sin(\theta-\alpha)°$ $(r > 0, 0° < \alpha° < 90°)$.

(ii) Hence solve the equation $5\sin\theta° - 4\cos\theta° = 3$, giving solutions to the nearest degree in the range $0°$ to $360°$.

(b) (i) Express $4\cos\theta° - 5\sin\theta°$ in the form $r\cos(\theta+\alpha)°$ $(r > 0$ and $0° < \alpha° < 90°)$.

(ii) Hence solve the equation $4\cos\theta° - 5\sin\theta° = -3$, giving solutions to the nearest degree in the range $0°$ to $360°$.

(c) Compare your answers to parts (a) and (b) and comment.

14 Solve the following equations for $0 < x < 2\pi$, giving solutions to two decimal places.

(a) $\sin x + 4\cos x = 3$ (b) $\sin x - 4\cos x = 3$ (c) $3\cos x - \sin x = 0.2$

(d) $\sin x + \sqrt{3}\cos x = 1$ (e) $\sqrt{2}\cos x + \sin x = 1$ (f) $2\sqrt{2}\cos x - \sin x = 2$

7 Natural logarithms and e^x

In this chapter you will learn
- what is meant by the number e
- how to find the natural logarithm of a number
- how to solve equations that involve e^x and ln x
- about the graphs of $y = e^x$ and $y = \ln x$

A Introducing e (answers p 187)

A1 You invest £1 in a savings scheme that gives 100% interest per annum!
How much money would you have at the end of one year?

A2 Now suppose that the interest is given as 50% compound interest twice a year.
Show that an investment of £1 will have grown to £2.25 at the end of the year.

A3 Now suppose that the interest is given as $33\frac{1}{3}$% compound interest three times a year.
Show that an investment of £1 will have grown to $£\left(1 + \frac{1}{3}\right)^3$ at the end of the year.
Work out this amount to the nearest penny.

Imagine this process continuing indefinitely.

A4 (a) Show that, if interest is calculated n times in one year year at a rate of $\dfrac{100}{n}$% each
time, then an investment of £1 will have grown to $£\left(1 + \dfrac{1}{n}\right)^n$ at the end of the year.

(b) Calculate the final amount if interest is added on 100 times a year.

(c) What happens to this final amount as n gets larger and larger?

A5 (a) Use a graph plotter to plot the graph of $y = \left(1 + \dfrac{1}{x}\right)^x$ for $x > 0$.

(b) Given that there is a limit for $\left(1 + \dfrac{1}{x}\right)^x$ as x tends to infinity, find this limit correct to
three decimal places.

In 1683, as part of some work on compound interest, the Swiss mathematician Jacob Bernouilli
tried to find the limit of $\left(1 + \dfrac{1}{n}\right)^n$ as n tends to infinity. He showed that the limit had to lie
between 2 and 3. Some years later, Leonhard Euler, another Swiss mathematician, used the
letter e to stand for this limit and gave an approximation for it to 18 decimal places as

$$e \approx 2.718\,281\,828\,459\,045\,235$$

This is the first time that a number was defined by a limiting process.

Euler probably chose e to stand for this number as the first four letters of the alphabet, a, b, c
and d, already had common mathematical uses. It is very unlikely that it has anything to do
with Euler's name: despite his great achievements in mathematics, he was a modest man.

The constant e is needed in various areas of mathematics, particularly in problems involving
growth and decay.

A6 Let f be a function defined for all non-negative integers by
$$f(n) = \frac{1}{0!} + \frac{1}{1!} + \frac{1}{2!} + \frac{1}{3!} + \ldots + \frac{1}{n!}.$$
(0! is conventionally defined as 1.)

(a) Show that $f(3) = 2\frac{2}{3}$.

(b) Evaluate these, correct to six decimal places.

    (i) $f(5)$ (ii) $f(8)$ (iii) $f(10)$ (iv) $f(15)$

(c) What do you notice about your results in part (b)?

In 1748, in *Introductio in analysin infinitorum*, Euler showed that
$$e = \frac{1}{0!} + \frac{1}{1!} + \frac{1}{2!} + \frac{1}{3!} + \ldots$$
and this is a much more efficient way of calculating the value of e.

Euler proved that e is irrational, so its decimal expansion never terminates or repeats. No matter how many digits in the expansion of e you know, the only way to find the next one is to calculate it using a definition such as the one above.

In general, the decimal expansion of e did not give rise to the same enthusiasm as that of π. However, by 1884 e had been calculated correct to 346 decimal places and in 1999 Sebastian Wedeniwski calculated e to 869 894 101 places!

Exercise A (answers p 188)

1 Without using a calculator, decide whether each statement is true or false.

(a) $e + 1 > 4$ (b) $\frac{e}{3} < 1$ (c) $2e > 5$ (d) $e^2 < 9$

(e) $\sqrt{e} < 2$ (f) $e^3 > 8$ (g) $\frac{1}{e} < 0$ (h) $e^{\frac{1}{3}} > 1$

(i) $e^\pi > 8$ (j) $e^{-2} > 0$ (k) $e^{-\frac{1}{2}} < 0$ (l) $e^0 = 1$

2 Use the e^x key on your calculator to evaluate these, correct to four decimal places.

(a) e^2 (b) e^{-3} (c) $\sqrt{e}$ (d) $e^{-\frac{1}{4}}$ (e) $\frac{1}{4}e$

3 Each of these is an approximation for e.
Put them in order, from the least accurate to the most accurate.

A $3 - \sqrt{\dfrac{5}{63}}$ B $\dfrac{271801}{99990}$ C $2 + \dfrac{54^2 + 41^2}{80^2}$ D $\left(\pi^4 + \pi^5\right)^{\frac{1}{6}}$

B Natural logarithms (answers p 188)

B1 Use trial and improvement to solve these equations, correct to two decimal places.

(a) $e^x = 7$ (b) $e^x = 2$

You know from earlier work on logarithms that, if a number y is expressed as $y = a^x$, then we say that x is the log-to-base-a of y and write it as $x = \log_a y$.

So, for example, if $2 = e^x$ then x is the log-to-base-e of 2 and can be written as $x = \log_e 2$.

K Logarithms to the base e have a special name and their own notation: they are called **natural logarithms** and we write $\log_e x$ as **ln x**.

In general, $e^x = y \Leftrightarrow x = \ln y$.

It can be useful to express this as $e^{\ln y} = y$ or $\ln e^x = x$.

The letters 'ln' are an abbreviation of the French 'logarithme naturel'.
The expression 'ln x' is usually pronounced as 'el-en-of-x' or 'lon-x'.

B2 (a) Find the ln key on your calculator and use it to calculate $\ln 2$ to six decimal places.

(b) Use your result to verify that $e^{\ln 2} = 2$.

B3 Without using a calculator, write down the value of $e^{\ln 4}$.

B4 (a) Use your calculator to evaluate $e^{1.3863}$ correct to three decimal places.

(b) Hence, without using a calculator, write down the value of $\ln 4$, correct to one decimal place.

B5 Solve these equations, correct to four decimal places.

(a) $e^x = 10$ (b) $e^x = 2.5$ (c) $e^x = 0.5$ (d) $e^x = 0.1$

B6 Explain why the equation $e^x = -5$ has no real solution.

B7 (a) Write $\dfrac{1}{e^3}$ in the form e^k for some integer k.

(b) Hence, without using a calculator, write down the value of $\ln \dfrac{1}{e^3}$.

B8 Without using a calculator, write down the value of these.

(a) $\ln e^5$ (b) $\ln e$ (c) $\ln \dfrac{1}{e^2}$ (d) $\ln \sqrt{e}$ (e) $\ln 1$

B9 Explain why a value for $\ln(-2)$ does not exist.

Of course, the laws of logarithms apply to natural logarithms.

K Where $a, b > 0$,
- $\ln a + \ln b = \ln ab$
- $\ln a - \ln b = \ln\left(\dfrac{a}{b}\right)$
- $\ln a^k = k \ln a$ (for all values of k)

B10 Show that $2\ln 3 + 5\ln 2$ is equivalent to the single logarithm $\ln 288$.

B11 Write each of these as a single logarithm.

(a) $\ln 3 + \ln 2$ (b) $\ln 8 - \ln 4$ (c) $2\ln 6$ (d) $2\ln 4 + 3\ln 2$

B12 (a) Use your calculator to find $\ln 5$ to six decimal places.

(b) Hence write down the value of each of these, correct to one decimal place.

(i) $\ln 25$ (ii) $\ln 5e$ (iii) $\ln \dfrac{e}{5}$

B13 Show that $\ln(8e^2) = 2 + 3\ln 2$.

You will see in the next chapter that there are many calculus problems where e^x and natural logarithms arise. This is one reason that e^x and $\ln x$ are such important functions.

In this chapter you will simplify expressions, solve equations and draw graphs where e and natural logarithms are involved. This is important preparation for working with e and natural logarithms in calculus.

Example 1

Solve the equation $e^{x+1} = 5$, giving your answer as an exact value.

Solution

$$e^{x+1} = 5$$
$$\Rightarrow \quad x + 1 = \ln 5$$
$$\Rightarrow \quad x = \ln 5 - 1$$

Example 2

Solve the equation $\ln x^3 = 7$, giving your answer correct to four decimal places.

Solution

$$\ln x^3 = 7$$
$$\Rightarrow \quad 3\ln x = 7$$
$$\Rightarrow \quad \ln x = \tfrac{7}{3}$$
$$\Rightarrow \quad x = e^{\frac{7}{3}}$$
$$\quad\quad = 10.3123 \text{ (to 4 d.p.)}$$

Example 3

Solve the equation $e^{-5x} - 3 = 0$, giving your answer correct to four decimal places.

Solution

$$e^{-5x} - 3 = 0$$
$$\Rightarrow \quad e^{-5x} = 3$$
$$\Rightarrow \quad -5x = \ln 3$$
$$\Rightarrow \quad x = -\tfrac{1}{5}\ln 3$$
$$\quad\quad = -0.2197 \text{ (to 4 d.p.)}$$

Example 4

Write the expression $4\ln(x+1) - \ln x$ as a single logarithm.

Solution

$$4\ln(x+1) - \ln x = \ln(x+1)^4 - \ln x$$
$$= \ln\frac{(x+1)^4}{x}$$

Example 5

Solve the equation $\ln(2x-1) + \ln 3 = 2$, giving your answer as an exact value.

Solution

$$\ln(2x-1) + \ln 3 = 2$$
$$\Rightarrow \qquad \ln(3(2x-1)) = 2$$
$$\Rightarrow \qquad \ln(6x-3) = 2$$
$$\Rightarrow \qquad 6x - 3 = e^2$$
$$\Rightarrow \qquad 6x = e^2 + 3$$
$$\Rightarrow \qquad x = \tfrac{1}{6}(e^2 + 3)$$

Exercise B (answers p 188)

1 Solve each equation, giving each solution correct to three decimal places.

(a) $e^x = 3$ (b) $4e^{-x} = 1$ (c) $e^x + 1 = 4.5$

(d) $e^{x+1} = 20$ (e) $e^{3x} = 0.2$ (f) $e^{\frac{1}{2}x} - 2 = 6$

(g) $2e^{4x} - 3 = 0$ (h) $e^{1-3x} = \tfrac{1}{2}$ (i) $e^{\frac{1}{4}x+1} - 1 = 0$

2 The growth of a colony of bacteria is modelled by the equation $y = 4e^t$, where t is measured in hours and y is the number of bacteria.

(a) Estimate the number of bacteria after

(i) 1 hour (ii) $3\tfrac{1}{2}$ hours (iii) 10 hours

(b) How long does it take for the colony to grow to 500 bacteria?

3 When a particular drug is injected into the body, the amount in the bloodstream is modelled by the equation $A = 5e^{-0.2t}$, where t is the time in hours after the dose is administered, and A is the amount remaining, in milligrams.

(a) What is the initial amount of drug in the bloodstream?

(b) How much of the drug (to the nearest 0.01 mg) remains in the blood after

(i) 1 hour (ii) 5 hours (iii) 24 hours

(c) How long does it take for the amount of drug to decrease to below 0.01 mg?

4 Show that $x = \tfrac{1}{3}(\ln 5 + 2)$ is the solution of the equation $e^{3x-2} = 5$.

5 Find the exact solution of each equation.

(a) $e^{2x+1} = 5$ (b) $e^{3x-4} = 1$ (c) $e^{1-\frac{1}{2}x} - 3 = 1$

6 Without using a calculator, show that   (a) $\ln\frac{1}{2} = -\ln 2$ (b) $e^{-\ln 4} = \frac{1}{4}$

7 Solve $e^{-2x} = \frac{1}{3}$ and write the solution in the form $x = a\ln b$, where b is a positive integer.

8 Find the exact solution of the equation $3e^{-5x} = \frac{1}{2}$.

9 Solve each equation, giving each solution correct to three decimal places.

(a) $\ln x = 4$ (b) $\ln 2x = -3$ (c) $3\ln x = 5$

(d) $\ln(x+1) = 0.6$ (e) $\ln(2x-3) = 1.5$ (f) $\ln\left(1 - \frac{1}{3}x\right) = -1.2$

10 Show that $x = \frac{1}{4}(e^2 + 1)$ is the solution of the equation $\ln(4x - 1) = 2$.

11 Find the exact solution of each equation.

(a) $\ln 4x = 5$ (b) $\ln(x-1) = -2$ (c) $\ln(2x+3) = 1$

12 Write each of these as a single logarithm.

(a) $3\ln x$ (b) $\ln 4x + \ln 5$ (c) $\ln\frac{1}{2}x + \ln 2x$

(d) $2\ln 5 + \ln x$ (e) $\ln x - \ln 2$ (f) $\frac{1}{2}\ln 9 - \ln x$

13 Find the exact solution of each equation.

(a) $\ln x + \ln 5 = 2$ (b) $\ln x - \ln 3 = 1$ (c) $\ln x + \ln 2x = 3$

(d) $3\ln 2 + \ln 5x = 0$ (e) $\ln(x+1) + \ln 2 = 5$ (f) $\ln(5-x) - \ln 4 = 1$

14 (a) Show that any solution to the equation $\ln(x+2) + \ln x = \ln 8$ must satisfy the equation $x^2 + 2x - 8 = 0$.

(b) Which of the solutions of $x^2 + 2x - 8 = 0$ is a solution of $\ln(x+2) + \ln x = \ln 8$? Explain your answer fully.

15 Solve each of these equations.

(a) $\ln(x+3) - \ln x = \ln 5$ (b) $\ln(x+1) + \ln x = \ln 6$

(c) $\ln(6x-5) = 2\ln x$ (d) $\ln(2x-1) + \ln 3x = 2\ln x$

16 (a) Show that the equation $e^{2x} - 4e^x + 3 = 0$ is equivalent to $y^2 - 4y + 3 = 0$ where $y = e^x$.

(b) Solve $e^{2x} - 4e^x + 3 = 0$, giving your solutions as exact values.

17 Solve each equation, giving all solutions as exact values.

(a) $e^{2x} - 6e^x + 5 = 0$ (b) $e^{2x} = 2e^x$ (c) $e^{2x} - 3e^x = 4$

18 Solve $2\ln(x+1) - \ln x = \ln 6$, giving your solutions as exact values.

19 Solve each equation, giving all solutions as exact values.

(a) $9e^x = 4e^{-x}$ (b) $e^{2x+1} = 3e^x$ (c) $5 - e^x = 4e^{-x}$

C Graphs (answers p 189)

The graph of $y = e^x$ is an exponential curve.

When $x = 0$ then $y = e^0 = 1$, so the y-intercept is 1.
As x gets very large and negative then e^x gets closer
and closer to 0 so the x-axis is an asymptote.

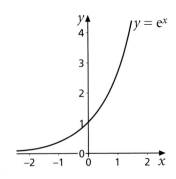

C1 Write down the range of the function $y = e^x$ where $x \in \mathbb{R}$.

We know that $y = e^x \Leftrightarrow x = \ln y$ so e^x and $\ln x$ are inverse functions and
the graph of $y = \ln x$ is the reflection of $y = e^x$ in the line $y = x$.

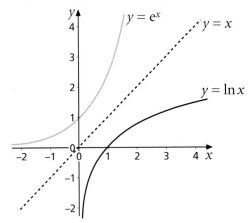

C2 Write down the largest suitable domain for the function $y = \ln x$.

C3 Sketch the graph of $y = f(x)$ for each function.

(a) $f(x) = e^{-x}, \ x \in \mathbb{R}$ (b) $f(x) = 2e^x, \ x \in \mathbb{R}$

(c) $f(x) = \ln(x + 1), \ x > -1$ (d) $f(x) = \ln(3x), \ x > 0$

C4 Write down the range of each function in C3.

Example 6

Sketch the graph of $y = e^{x-1} - 3$, showing clearly the exact coordinates of
any points where the graph meets the axes.

Solution

The graph of $y = e^{x-1} - 3$ is obtained from $y = e^x$ by a translation of 1 unit to the right and
3 units down. So $y = -3$ is an asymptote of the graph of $y = e^{x-1} - 3$.

When $x = 0$, $y = e^{-1} - 3 = \dfrac{1}{e} - 3$ so the graph cuts the y-axis at $\left(0, \dfrac{1}{e} - 3\right)$.

When $y = 0$, $e^{x-1} - 3 = 0$

$\Rightarrow \quad e^{x-1} = 3$

$\Rightarrow \quad x - 1 = \ln 3$

$\Rightarrow \quad x = \ln 3 + 1$

so the graph cuts the x-axis at $(\ln 3 + 1, 0)$.

Hence a sketch is:

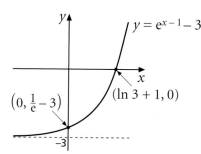

The asymptote is shown by a dotted line.

Finding approximate decimal values for any important coordinates helps you draw a realistic sketch:

$\left(0, \dfrac{1}{e} - 3\right) \approx (0, -2.6)$ *and*

$(\ln 3 + 1, 0) \approx (2.1, 0)$

Example 7

A function is defined by the rule $f(x) = \ln(2x + 5)$.

Sketch the graph of $y = f(x)$, showing clearly the coordinates of any points where the graph meets the axes.

Write down the largest suitable domain for the function f.

Solution

$y = \ln(2x + 5)$ *is a transformation of* $y = \ln x$ *so we know the general shape.*

When $x = 0$, $y = \ln 5$ so the graph cuts the y-axis at $(0, \ln 5)$.

When $y = 0$, $\ln(2x + 5) = 0$

$\Rightarrow \quad 2x + 5 = 1$

$\Rightarrow \quad x = -2$

so the graph cuts the x-axis at $(-2, 0)$.

$x = 0$ *is an asymptote to* $y = \ln x$ *so* $2x + 5 = 0$ *is an asymptote to* $y = \ln(2x + 5)$.

$2x + 5 = 0$ when $x = -2\frac{1}{2}$ so the line $x = -2\frac{1}{2}$ is a vertical asymptote.

Hence a sketch is:

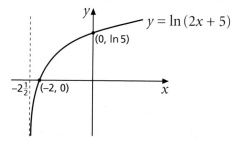

The asymptote $x = -2\frac{1}{2}$ is shown by a dotted line.

The largest suitable domain is $x > -2\frac{1}{2}$.

Example 8

On the same axes, sketch the graphs of $y = e^{3x}$ and $y = 3e^{-x}$.
Find the x-coordinate of the point of intersection, correct to four significant figures.

Solution

The graph of $y = e^{3x}$ is obtained from $y = e^x$ by a stretch of factor $\frac{1}{3}$ in the x-direction.

The graph of $y = 3e^{-x}$ is obtained from $y = e^x$ by a stretch of factor 3 in the y-direction followed by reflection in the y-axis (or vice versa).

Hence a sketch is:

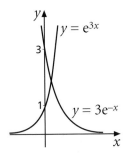

The x-coordinate of the point of intersection is the solution of the equation $e^{3x} = 3e^{-x}$.

$$e^{3x} = 3e^{-x} \quad \Rightarrow \quad e^{3x} = \frac{3}{e^x}$$

$$\Rightarrow \quad e^{3x} \times e^x = \frac{3}{e^x} \times e^x$$

$$\Rightarrow \quad e^{4x} = 3$$

$$\Rightarrow \quad 4x = \ln 3$$

$$\Rightarrow \quad x = \tfrac{1}{4}\ln 3 = 0.2747 \text{ (to 4 s.f.)}$$

Exercise C (answers p 189)

1 A function is defined by $f(x) = e^x + 5, \; x \in \mathbb{R}$.

(a) Evaluate $f(0)$.

(b) Sketch the graph of $y = f(x)$, indicating the position of the horizontal asymptote.

(c) What is the range of f?

2 Sketch the graph of $y = f(x)$ where $f(x) = e^{x+5}, \; x \in \mathbb{R}$.
Indicate clearly the exact coordinates of the point where the graph cuts the y-axis.

3 What transformation will map the graph of $y = e^x$ on to the graph of $y = e^{\frac{1}{3}x}$?

4 (a) What transformation will map the graph of $y = \ln x$ on to the graph of $y = \ln x - 1$?

(b) Show that the graph of $y = \ln x - 1$ meets the x-axis at the point $(e, 0)$.

5 For each function, sketch the graph of $y = f(x)$, indicating clearly the exact coordinates of any points where the graph meets the x-axis.

(a) $f(x) = \ln x - 3, \; x > 0$ (b) $f(x) = 3\ln x, \; x > 0$

(c) $f(x) = -\ln x, \; x > 0$ (d) $f(x) = 1 - \ln x, \; x > 0$

6 Describe a sequence of geometrical transformations by which the graph of $y = 5e^{-x}$ can be obtained from that of $y = e^x$.

7 Sketch the graph of $y = e^{-x} - 4$, indicating clearly the exact coordinates of any points where the graph meets the axes.

8 A function is defined by $f(x) = e^{x+2} - 5$, $x \in \mathbb{R}$.

   (a) Evaluate $f(-2)$.

   (b) Sketch the graph of $y = f(x)$, indicating clearly the exact coordinates of any points where the graph meets the axes.

   (c) What is the equation of the asymptote?

   (d) What is the range of f?

   (e) Solve the equation $f(x) = 1$.

9 **(a)** What transformation will map the graph of $y = \ln x$ on to the graph of $y = \ln(x + 2)$?

   (b) Sketch the graph of $y = \ln(x + 2)$, showing clearly the exact coordinates of any points where the graph meets the axes and the position of the vertical asymptote.

10 A function is defined by $f: x \mapsto 2e^{-x} - 1$, $x \in \mathbb{R}$. A sketch of $y = f(x)$ is shown.

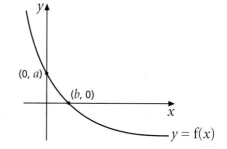

   (a) What is the value of a?

   (b) Show that $b = \ln 2$.

   (c) What is the range of f?

   (d) Solve the equation $f(x) = 3$.

11 The rule for a function is $g: x \mapsto 3 + \ln(2x + 1)$. A sketch of $y = g(x)$ is shown.

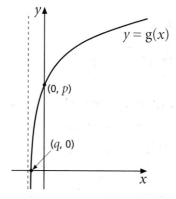

   (a) What are the values of p and q?

   (b) What is the equation of the vertical asymptote?

   (c) Solve the equation $g(x) = -2$.

12 A function is defined by $f(x) = 5 - 2\ln(x + 3)$, $x > -3$. Sketch the graph of $y = f(x)$.

13 **(a)** On the same axes, sketch the graphs of $y = e^{-x}$ and $y = 3e^x$.

   (b) The two curves $y = e^{-x}$ and $y = 3e^x$ intersect at the point P.

     (i) Show that the x-coordinate of the point P is a root of the equation $e^{2x} = \frac{1}{3}$.

     (ii) Find the x-coordinate of the point P, correct to four significant figures.

14 A cup of coffee, initially at boiling point, cools according to Newton's law of cooling, so that after t minutes its temperature, $T°C$, is given by

$$T = 15 + 85e^{-\frac{t}{8}}$$

(a) Sketch the graph of T against t.

(b) How long does the coffee take to cool to $40°C$?

15 A function is defined by $f(x) = |e^x - 3|$, $x \in \mathbb{R}$.

(a) Evaluate $f(0)$.

(b) Evaluate these, correct to two decimal places.

(i) $f(2)$ (ii) $f(1)$ (iii) $f(-3)$

(c) Solve the equation $f(x) = 0$.

(d) Sketch the graph of $y = f(x)$, showing clearly the exact coordinates of any points where the graph meets the axes.

(e) What is the range of f?

(f) Show that the equation $f(x) = 1$ has two solutions.
Solve $f(x) = 1$, giving the solutions in exact form.

(g) Solve $f(x) = 5$.

16 A function is defined by $g(x) = \ln|x - 2|$, $x \neq 2$.

(a) Evaluate $g(3)$ and $g(1)$.

(b) Sketch the graph of $y = g(x)$, showing clearly the exact coordinates of any points where the graph meets the axes.

(c) Solve $g(x) = 3$, giving the solutions in exact form.

D Inverses

You have seen that one way to find a rule for the inverse of a function is to write the function in terms of x and y and then rearrange to obtain a rule for x in terms of y.

For example, let f be the function defined by $\quad f(x) = e^x + 1$, $x \in \mathbb{R}$.

To find the inverse first write the rule as $\qquad y = e^x + 1$

Then rearrange to obtain $\qquad\qquad\qquad y - 1 = e^x$

$$\Rightarrow \quad \ln(y - 1) = x$$

So the inverse rule is $f^{-1}(x) = \ln(x - 1)$.

The range of f is the set of real numbers given by $f(x) > 1$ so the domain of f^{-1} is $x > 1$.
Hence we can define the inverse fully by

$$f^{-1}(x) = \ln(x - 1), \ x > 1$$

The range of f^{-1} is $f^{-1}(x) \in \mathbb{R}$ (the domain of f).

You have also seen that reflecting the graph of a function in the line $y = x$ gives the graph of its inverse.

The graphs of $y = f(x)$ and $y = f^{-1}(x)$ are:

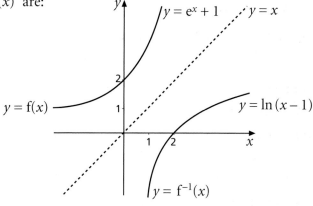

Example 9

The function f is defined for $x > 0$ by $f(x) = \ln(3x) - 6$.
Find an expression for $f^{-1}(x)$.

Solution

First write the rule as $\qquad\qquad y = \ln(3x) - 6$

Rearrange. $\qquad\qquad\qquad y + 6 = \ln(3x)$

$\qquad\Rightarrow\qquad\qquad e^{y+6} = 3x$

$\qquad\Rightarrow\qquad\qquad \frac{1}{3}e^{y+6} = x$

So $f^{-1}(x) = \frac{1}{3}e^{x+6}$.

Example 10

The function g is defined for $x \in \mathbb{R}$ by $g(x) = 2e^{x-5} + 3$.
Find an expression for $g^{-1}(x)$.
State the domain and range of g^{-1}.

Solution

First write the rule as $\qquad\qquad\qquad y = 2e^{x-5} + 3$

Rearrange. $\qquad\qquad\qquad\qquad y - 3 = 2e^{x-5}$

$\qquad\Rightarrow\qquad\qquad \frac{1}{2}(y - 3) = e^{x-5}$

$\qquad\Rightarrow\qquad \ln\left(\frac{1}{2}(y-3)\right) = x - 5$

$\qquad\Rightarrow\ \ln\left(\frac{1}{2}(y-3)\right) + 5 = x$

So $g^{-1}(x) = \ln\left(\frac{1}{2}(x-3)\right) + 5$.

The range of g is $g(x) > 3$ so the domain of g^{-1} is $x > 3$.
The domain of g is $x \in \mathbb{R}$ so the range of g^{-1} is $g^{-1}(x) \in \mathbb{R}$.

Exercise D (answers p 191)

1 For each rule, find an expression for $f^{-1}(x)$, where f^{-1} is the inverse of f.

(a) $f(x) = e^{2x}$ (b) $f(x) = e^{x+2}$ (c) $f(x) = e^x + 2$

(d) $f(x) = 2e^x$ (e) $f(x) = e^{2-x}$ (f) $f(x) = e^{-x} - 2$

2 For each rule, find an expression for $g^{-1}(x)$, where g^{-1} is the inverse of g.

(a) $g(x) = \ln(5x)$ (b) $g(x) = \ln(x+5)$ (c) $g(x) = 5 + \ln x$

(d) $g(x) = 5\ln x$ (e) $g(x) = \ln(\frac{1}{5}x)$ (f) $g(x) = 5\ln x - 1$

3 A function is defined by $f(x) = e^{x-1} + 2$, $x \in \mathbb{R}$.

(a) What is the range of f?

(b) Find an expression for $f^{-1}(x)$.

(c) State the domain and range of f^{-1}.

(d) On the same axes, sketch the graphs of $y = f(x)$ and $y = f^{-1}(x)$.
Show clearly any asymptotes and state their equations.

4 A function is defined by $g(x) = \ln(x+1) - 2$, $x > -1$.

(a) (i) Find an expression for $g^{-1}(x)$.

(ii) Hence find the exact solution to the equation $g(x) = 7$.

(b) State the domain and range of g^{-1}.

5 A function is defined by $f(x) = 2e^x + 3$, $x \in \mathbb{R}$.

(a) Find an expression for $f^{-1}(x)$.

(b) (i) State the domain of f^{-1}.

(ii) Hence show that the equation $f(x) = 1$ has no solution.

6 A function is defined by $h(x) = \ln(5x - 10)$, $x > 2$.

(a) Find an expression for $h^{-1}(x)$.

(b) On the same axes, sketch the graphs of $y = h^{-1}(x)$ and $y = h(x)$.
Show clearly any asymptotes and state their equations.

7 A function is defined by $f(x) = 3e^{-x}$, $x \geq -2$.

(a) What is the maximum value of $f(x)$?

(b) Find an expression for $f^{-1}(x)$.

(c) What is the domain and range of f^{-1}?

8 A particular type of car depreciates in value according to the mathematical model

$$V = 14\,000\,e^{-0.3t}$$

where £V is the value t years after it is sold as new.

(a) Find a formula for t in terms of V.

(b) Sarah buys one of these cars as new at the beginning of the year 2005.
During which year does the value of the car fall to below £5000?

- e is an irrational number; its value is 2.718 281 828 correct to nine decimal places. (pp 96–97)

- Natural logarithms use e as a base and we write $\log_e x$ as $\ln x$:
$$e^x = y \iff x = \ln y$$ (p 98)

- $y = \ln x$ and $y = e^x$ are inverse functions so the graph of $y = \ln x$ is the reflection of $y = e^x$ in the line $y = x$.

  e^x is positive for all x so $\ln x$ is defined only for positive values of x.

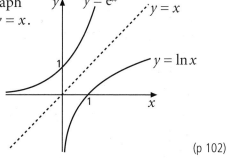

(p 102)

Mixed questions (answers p 191)

1 (a) The diagram shows the graph of $y = f(x)$, where the function f is defined by

$$f(x) = 40e^{\frac{x}{50}}, \ x \in \mathbb{R}$$

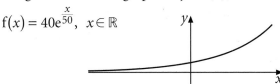

   (i) Write down the coordinates of the point where the graph intersects the y-axis.

   (ii) State the range of the function f.

   (iii) Express $f(\ln 10)$ in the form 40×10^P where p is a constant to be found.

(b) The function g is defined by

$$g(x) = 40\left(25 + e^{\frac{x}{50}} \right), \ x \in \mathbb{R}$$

   (i) State the range of the function g.

   (ii) Sketch the graph of $y = g(x)$.

   (iii) Show that $g(x) = 1200 \implies x = 50 \ln 5$.

(c) The number of bacteria in a particular culture is discovered to be $40\left(25 + e^{\frac{t}{50}} \right)$ where t is the time in days from the start of the experiment.

   (i) State the number of bacteria in the culture at the start of the experiment.

   (ii) Calculate the time it takes for the number of bacteria to reach 1200. Give your answer to the nearest tenth of a day.

2 A function is defined by $f(x) = 1 - 3e^{-x}, \ x \in \mathbb{R}$.

   (a) Sketch the graph of $y = f(x)$.   **(b)** Solve the inequality $f(x) > 0$.

3 The functions f and g are defined by $f: x \mapsto \ln\left(\frac{1}{2}x - 3\right), \ x > 6$

$\qquad\qquad\qquad\qquad\qquad\qquad\quad g: x \mapsto 4e^{2x}, \ x \in \mathbb{R}$

   (a) Find an expression for $f^{-1}(x)$.

   (b) Evaluate $gf(8)$.

   (c) Show that $gf(x) = (x - 6)^2$.

4 The functions f and g are defined by $f(x) = 1 + \ln x, \ x > 0$

$\qquad\qquad\qquad\qquad\qquad\qquad\quad g(x) = (ex)^2, \ x \in \mathbb{R}$

   (a) Evaluate $g(5)$ correct to two decimal places.

   (b) Describe the geometrical transformation by which the graph of $y = f(x)$ can be obtained from the graph of $y = \ln x$.

   (c) Find an expression for $f^{-1}(x)$.

   (d) Show that $fg(x) = 3 + 2\ln x$.

   (e) Sketch the graph of $y = |fg(x)|$.

5 The function f is given by $f: x \mapsto \ln(3x - 9), \ x > 3$.

   (a) Find, to three significant figures, the value of x for which $f(x) = 5$.

   The function g is given by $g: x \mapsto \ln|3x - 9|, \ x \neq 3$.

   (b) Sketch the graph of $y = g(x)$.

Test yourself (answers p 192)

1 Write each of these expressions in the form $a \ln p + b \ln q$.

   (a) $\ln(pq)$ **(b)** $\ln(p^3 q^2)$ **(c)** $\ln\left(\dfrac{p^2}{q}\right)$ **(d)** $\ln\sqrt{\dfrac{p^2}{q}}$

2 The functions f and g are defined for all real numbers by

$\qquad f: x \mapsto e^{4x}$

$\qquad g: x \mapsto \frac{1}{4}x - 3$

   (a) Evaluate $fg(12)$.

   (b) Find an expression for $f^{-1}(x)$.

   (c) Sketch on the same diagram the graphs of $y = f(x)$ and $y = f^{-1}(x)$, indicating the relationship between the graphs.

   (d) Sketch the graph of $y = |gf(x)|$, indicating clearly the exact coordinates of any points at which the graph meets the axes.

3 Describe, in each of the following cases, a single transformation which maps the graph of $y = e^x$ on to the graph of the function given.

   (a) $y = e^{4x}$ **(b)** $y = e^{x+1}$ **(c)** $y = -e^x$

4 Find the exact solution of the equation $\ln(3y + 2) - \ln 5 = 2$.

5 The function f is defined by $f(x) = e^{x-3} + 2$, $x \in \mathbb{R}$.

   (a) Sketch the graph of $y = f(x)$, showing clearly the coordinates of any points where the graph meets the axes.

   (b) What is the range of f?

   (c) Solve the equation $f(x) = 7$.

   (d) Find an expression for $f^{-1}(x)$ and state the domain of f^{-1}.

6 The function f is defined for $x > 0$ by $f(x) = 3 \ln x$.

   (a) Describe the geometrical transformation by which the graph of $y = f(x)$ can be obtained from the graph of $y = \ln x$.

   (b) Find an expression for $f^{-1}(x)$.

7 The diagram shows the graph of $y = f(x)$, where f is defined by $f(x) = \ln(2x + 7)$, $x > -3\frac{1}{2}$.

The line $x = a$ is an asymptote. The points $(b, 0)$ and $(0, c)$ are the x- and y-intercepts respectively.

Find the exact values of a, b and c.

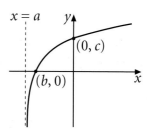

8 The function f is given by

$$f: x \mapsto \ln(4 - 2x), \ x \in \mathbb{R}, \ x < 2$$

   (a) Find an expression for $f^{-1}(x)$.

   (b) Sketch the curve with equation $y = f^{-1}(x)$, showing the coordinates of the points where the curve meets the axes.

The function g is given by

$$g: x \mapsto 3^x, \ x \in \mathbb{R}$$

   (c) Find the value of x for which $g(x) = 1.5$, giving your answer to three decimal places.

   (d) Evaluate $gf(1)$ to three decimal places.

Edexcel

9 The diagram shows the graph of $y = f(x)$, where f is defined by

$$f(x) = 3e^{-x}, \ x \in \mathbb{R}$$

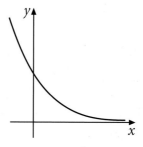

   (a) Describe a sequence of geometrical transformations by which the above graph can be obtained from the graph of $y = e^x$.

   (b) Solve the equation $f(x) = 0.1$, giving your answer correct to two decimal places.

   (c) Find an expression for $f^{-1}(x)$.

   (d) State the domain and range of f^{-1}.

   (e) Sketch the graph of $y = f(|x|)$.

8 Differentiation

In this chapter you will learn how to
- differentiate e^x, $\ln x$, $\sin x$, $\cos x$ and $\tan x$
- differentiate a product of two functions, a quotient of two functions and a function of a function

Key points from Core 1 and Core 2

- The derivative of x^n is nx^{n-1}.

- The derivative of $f(x) + g(x)$ is $f'(x) + g'(x)$.
  The derivative of $kf(x)$ is $kf'(x)$.

- If $f'(a) > 0$, f is increasing at $x = a$. If $f'(a) < 0$, f is decreasing at $x = a$.

- Points where $f'(x) = 0$ are called stationary points of the function f.
  If $f'(a) = 0$ and $f''(a) > 0$, then $x = a$ is a local minimum.
  If $f'(a) = 0$ and $f''(a) < 0$, then $x = a$ is a local maximum.

A Exponential functions (answers p 193)

The functions $y = 2^x$ and $y = 3^x$ are exponential functions.
Their graphs are shown here.

Because 2^0 and 3^0 are both 1, both graphs go through $(0, 1)$.

In fact, for any value of a, the graph of $y = a^x$ goes through $(0, 1)$.

At $(0, 1)$, the graph of $y = 3^x$ is steeper than the graph of $y = 2^x$.

We shall first find the gradient of each graph at $(0, 1)$.

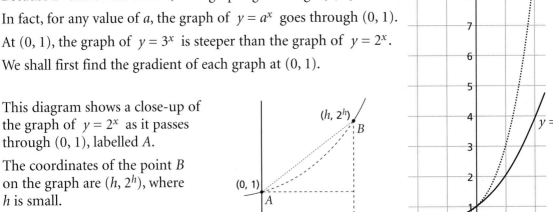

This diagram shows a close-up of the graph of $y = 2^x$ as it passes through $(0, 1)$, labelled A.

The coordinates of the point B on the graph are $(h, 2^h)$, where h is small.

A1 (a) Show that the gradient
   of the line AB is $\dfrac{2^h - 1}{h}$.

   (b) Find the value of this
   expression when $h = 0.1$.

   (c) Do the same for $h = 0.05$, 0.01, 0.001 and 0.0001.

   (d) What conclusion do you draw about the gradient of $y = 2^x$ at the point $(0, 1)$?

A2 If the graph of $y = 2^x$ is replaced by the graph of $y = 3^x$, the coordinates of B become $(h, 3^h)$ and the gradient of AB is $\dfrac{3^h - 1}{h}$.

Find the value of this expression when $h = 0.1, \ 0.01, \ 0.001$ and 0.0001 and hence find an approximate value for the gradient of $y = 3^x$ at the point $(0, 1)$.

You should find that at the point $(0, 1)$ the gradient of $y = 2^x$ is approximately 0.69 and the gradient of $y = 3^x$ is approximately 1.1.

This suggests that somewhere between 2 and 3 is a value of a for which the gradient of $y = a^x$ at the point $(0, 1)$ is equal to 1.

We would expect this value to be nearer to 3 than to 2, because 1.1 is nearer to 1 than is 0.69. You could use a spreadsheet to find this value by trial and improvement.

It can be shown, by methods that are beyond the scope of this book, that the required number has the value $2.7182818\ldots$, which is the number denoted by e.

A3 Verify by using a calculator that when you substitute $h = 0.1, 0.01, 0.001, \ldots$ into the expression $\dfrac{e^h - 1}{h}$, the results get closer and closer to 1.

The graph of $y = e^x$ is shown on the right. Its gradient at the point $(0, 1)$ is 1.

We now need to investigate the gradient at any other point of the graph, say the point P where $x = p$.

This diagram is a close-up of the graph as it passes through P, whose coordinates are (p, e^p).

Q is a nearby point on the graph, where $x = p + h$ and h is small. The coordinates of Q are $(p + h, e^{p + h})$.

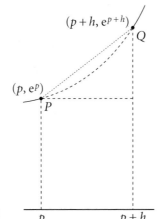

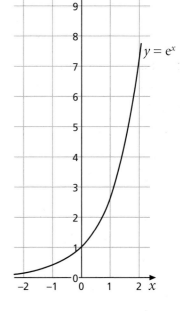

The gradient of the line PQ is
$$\frac{e^{p + h} - e^p}{h}$$

A4 (a) Show that this expression can be written as
$$\frac{e^p(e^h - 1)}{h}$$

(b) Explain why, as h gets smaller and smaller, the value of this expression gets closer and closer to e^p. What conclusion do you draw about the gradient of $y = e^x$ at the point where $x = p$?

The result of A4 (b) is this: the gradient of the curve $y = e^x$ when $x = p$ is e^p.
This can be stated simply as:

K The derivative of e^x is e^x.

This is the main reason the number e is important in mathematics.

A5 Write down the derivative of each of these.

(a) $3e^x + 4$ (b) $5e^x + 4x$ (c) $2e^x - x^2$ (d) $e^x + \dfrac{2}{x}$

A6 Use the fact that $e^{x+2} = e^2 e^x$ to find the derivative of e^{x+2}.
(Remember that e^2 is just a number.)

Example 1

Given that $f(x) = 3e^x - \dfrac{1}{x}$, show that f(x) is increasing for $x > 0$.

Solution

$f(x) = 3e^x - x^{-1}$

$f'(x) = 3e^x - (-1)x^{-2} = 3e^x + \dfrac{1}{x^2}$. When $x > 0$, $e^x > 0$ and $\dfrac{1}{x^2} > 0$, so $3e^x + \dfrac{1}{x^2} > 0$.

So f(x) is increasing for $x > 0$.

Exercise A (answers p 193)

1 Find the derivative of each of these.

(a) $e^x - 5x$ (b) $x^2 + 4e^x$ (c) $e^x - 3\sqrt{x}$ (d) $3e^x + x^3 - 1$

2 Find $f'(x)$ given that

(a) $f(x) = 2e^x - x^{-2}$ (b) $f(x) = 6e^x + \dfrac{1}{x}$ (c) $f(x) = x^{\frac{3}{2}} + 2e^x$ (d) $f(x) = \dfrac{1}{\sqrt{x}} + 2e^x$

3 (a) Find $\dfrac{dy}{dx}$ given that $y = e^x - 3x$.

(b) Show that $\dfrac{dy}{dx} = 0$ when $x = \ln 3 = 1.099$ to 3 d.p.

(c) From (b) it follows that the graph of $y = e^x - 3x$ has a stationary point where $x = 1.099$. Find the y-coordinate of this stationary point, to 3 d.p.

(d) By differentiating $\dfrac{dy}{dx}$ find $\dfrac{d^2y}{dx^2}$ for the function $y = e^x - 3x$.

(e) Find the value of $\dfrac{d^2y}{dx^2}$ at the stationary point.

(f) What type of stationary point is it? Give the reason for your answer.

4 Show that the function f given by $f(x) = e^x - \dfrac{1}{x^3}$ is increasing for $x > 0$.

B The derivative of ln x (answers p 194)

Because $\ln x$ is the inverse of e^x, the graph of $y = \ln x$ is the reflection of $y = e^x$ in the line $y = x$.

P is the point on the graph of $y = \ln x$ where $x = 3$. So P is $(3, \ln 3)$.

The corresponding point Q on $y = e^x$ is $(\ln 3, 3)$.

On the graph of $y = e^x$ the gradient is equal to the value of y, which is 3 at Q.

From the diagram, the gradient of $y = \ln x$ at P is the reciprocal of 3, which is $\frac{1}{3}$.

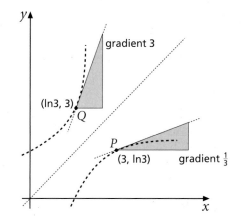

There is nothing special about the number 3 here.

For all positive values of x, the derivative of $\ln x$ is $\frac{1}{x}$.

This can be shown more formally. But first we need a more general result.

The definition of $\frac{dy}{dx}$ involves making small changes δx and δy to x and y and seeing what happens to the ratio $\frac{\delta y}{\delta x}$ as δx and δy get smaller and smaller.

We can also look at the ratio $\frac{\delta x}{\delta y}$, which is equal to $\frac{1}{\frac{\delta y}{\delta x}}$.

The limiting value of $\frac{\delta x}{\delta y}$ as δx and δy get smaller and smaller is the derivative of x with respect to y. It is denoted by $\frac{dx}{dy}$, which is equal to $\frac{1}{\frac{dy}{dx}}$.

It follows that

$$\frac{dy}{dx} = \frac{1}{\frac{dx}{dy}}$$

This relationship helps us to differentiate the function $\ln x$, which is the inverse of e^x.

Let $y = \ln x$

This can be rewritten as $\qquad x = e^y$

Differentiate with respect to y: $\quad \dfrac{dx}{dy} = e^y$

Use the relationship above: $\quad \dfrac{dy}{dx} = \dfrac{1}{\frac{dx}{dy}} = \dfrac{1}{e^y} = \dfrac{1}{x}$

The derivative of $\ln x$ is $\frac{1}{x}$.

B1 Differentiate each of these with respect to x.

(a) $\ln x + x^2$ (b) $\ln x - \sqrt{x}$ (c) $\ln x + 5e^x$ (d) $\dfrac{1}{x} - \ln x$

B2 Explain why $\ln 3x = \ln 3 + \ln x$ and hence explain why the derivative of $\ln 3x$ is $\dfrac{1}{x}$.

B3 (a) Explain why $\ln(5x^2) = \ln 5 + 2\ln x$ and hence differentiate $\ln(5x^2)$.

(b) Differentiate   **(i)** $\ln(2x^3)$ **(ii)** $\ln(\sqrt{x})$ **(iii)** $\ln(x^{-3})$ **(iv)** $\ln\left(\dfrac{1}{x}\right)$

Example 2

Find $f'(x)$ given that $f(x) = 5x^4 - \ln(5x^4)$.

Solution

$f(x) = 5x^4 - (\ln 5 + 4\ln x) = 5x^4 - \ln 5 - 4\ln x$

$$f'(x) = 20x^3 - 4\left(\frac{1}{x}\right) = 20x^3 - \frac{4}{x} \quad (\ln 5 \text{ is a constant, so its derivative is } 0.)$$

Example 3

Find the stationary point on the graph of $y = 8x^2 - \ln x$ and determine its type.

Solution

$\dfrac{dy}{dx} = 16x - \dfrac{1}{x}$. At a stationary point, $16x - \dfrac{1}{x} = 0 \Rightarrow 16x^2 - 1 = 0 \Rightarrow x = \pm\frac{1}{4}$.

When $x = \frac{1}{4}$, $y = \frac{8}{16} - \ln\frac{1}{4} = 1.886$ (to 3 d.p.)

When $x = -\frac{1}{4}$, there is no value of y because $\ln\left(-\frac{1}{4}\right)$ does not exist.

So the stationary point is $(0.25, 1.886)$.

$\dfrac{d^2y}{dx^2} = 16 + \dfrac{1}{x^2}$. When $x = \frac{1}{4}$, $\dfrac{d^2y}{dx^2} = 16 + 16 = 32$. As $32 > 0$, the point is a minimum.

Exercise B (answers p 194)

1 Differentiate these with respect to x.

(a) $x^2 + \ln x$ (b) $3\ln x - x^3$ (c) $\sqrt{x} - 2\ln x$ (d) $\ln x + e^x$

2 Express $\ln(7x^3)$ in the form $\ln a + b\ln x$ and hence differentiate $\ln(7x^3)$.

3 Find $f'(x)$ given that

(a) $f(x) = \ln(6x^2)$ (b) $f(x) = \ln(4x^5)$ (c) $f(x) = \ln(2\sqrt{x})$ (d) $f(x) = \ln\left(3x^{\frac{3}{2}}\right)$

4 Express $\ln\left(\dfrac{3}{x^2}\right)$ in the form $\ln a - b\ln x$ and hence differentiate $\ln\left(\dfrac{3}{x^2}\right)$.

5 Find $\dfrac{dy}{dx}$ given that

(a) $y = e^x + \ln(2x^3)$ (b) $y = \ln\left(\dfrac{3}{x}\right)$ (c) $y = \ln\left(x^{\frac{2}{3}}\right)$ (d) $y = x^{-\frac{1}{2}} + \ln\left(x^{-\frac{1}{2}}\right)$

6 (a) Show that the graph of $y = \ln x - x$ has a stationary point at $x = 1$.

(b) Show that the stationary point is a maximum.

7 Find the stationary point on the graph of $y = x^4 - \frac{1}{4}\ln x$ and determine its type.

C Differentiating a product of functions (answers p 194)

Suppose that the variables u and v are defined in terms of x by

$$u = x^2 \qquad v = 2x + 1$$

Let $y = uv$.

It is a simple matter to find $\dfrac{dy}{dx}$ in this case, because $y = x^2(2x + 1) = 2x^3 + x^2$.

C1 (a) Find $\dfrac{dy}{dx}$.

(b) y is the product of u and v. Show that $\dfrac{dy}{dx}$ is **not** the product of $\dfrac{du}{dx}$ and $\dfrac{dv}{dx}$.

You can't differentiate the product of two functions by differentiating each one separately and multiplying the results. However, there is a rule for differentiating a product, and this is what we shall now investigate.

Let u and v be two variables that are functions of x.
The product uv represents the area y of a rectangle.

Suppose that x increases by a small amount δx.

As a result, u increases by a small amount δu, and v by a small amount δv, and the area y by a small amount δy.

These diagrams show the rectangle before and after the increases.

The second rectangle can be split up like this:

C2 Show that $\delta y = u\,\delta v + v\,\delta u + \delta u\,\delta v$.

When the equation above is divided throughout by δx it becomes

$$\frac{\delta y}{\delta x} = u\frac{\delta v}{\delta x} + v\frac{\delta u}{\delta x} + \frac{\delta u}{\delta x}\delta v$$

Now let δx get smaller and smaller.

$\dfrac{\delta y}{\delta x}$ gets closer and closer to $\dfrac{dy}{dx}$. $\dfrac{\delta u}{\delta x}$ gets closer and closer to $\dfrac{du}{dx}$.

$\dfrac{\delta v}{\delta x}$ gets closer and closer to $\dfrac{dv}{dx}$. δv gets closer and closer to 0.

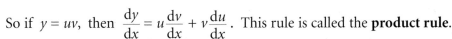

So if $y = uv$, then $\dfrac{dy}{dx} = u\dfrac{dv}{dx} + v\dfrac{du}{dx}$. This rule is called the **product rule**.

The rule can also be stated using function notation:

The derivative of $f(x)g(x)$ is $f'(x)g(x) + f(x)g'(x)$.

C3 Let $u = x^2$ and $v = 2x + 1$, as in question C1.
Use the product rule to find the derivative of uv.
Check that the result agrees with the previous one.

C4 Let $f(x) = x^3$ and $g(x) = x^2 - 3$.

(a) Use the product rule to find the derivative of $f(x)g(x)$.

(b) Find the derivative of $f(x)g(x)$ in a different way by first multiplying $f(x)$ and $g(x)$ and then differentiating. Check that the two results agree.

C5 Use the product rule to find the derivative of $x^3 e^x$.

C6 Use the product rule to find $\dfrac{dy}{dx}$ given that $y = x^4 \ln x$.

C7 Differentiate these with respect to x.

(a) $(x^2 - 3x)e^x$ (b) $(x^2 - 3x)\ln x$ (c) $x^{\frac{1}{2}} \ln x$

Example 4

(a) Find the exact values of the coordinates of the stationary points on the graph of $y = x^2 e^x$.

(b) Determine the type of each stationary point.

Solution

(a) Use the product rule with $u = x^2$ and $v = e^x$.

$$\frac{du}{dx} = 2x, \ \frac{dv}{dx} = e^x, \ \text{so} \ \frac{dy}{dx} = u\frac{dv}{dx} + v\frac{du}{dx} = x^2 e^x + e^x(2x) = (x^2 + 2x)e^x$$

At stationary points, $\dfrac{dy}{dx} = 0$

$\Rightarrow \quad (x^2 + 2x)e^x = 0$

$\Rightarrow \quad x(x + 2)e^x = 0 \quad\quad$ (*factorising* $x^2 + 2x$)

$\Rightarrow \quad\quad x = 0 \ \text{or} \ x = -2 \quad$ (because e^x is never 0)

When $x = 0$, $y = 0^2 e^0 = 0$. When $x = -2$, $y = (-2)^2 e^{-2} = 4e^{-2}$.

So the stationary points are $(0, 0)$ and $(-2, 4e^{-2})$.

(b) *To find* $\dfrac{d^2 y}{dx^2}$, *you have to differentiate* $\dfrac{dy}{dx}$, *which is* $(x^2 + 2x)e^x$.

Use the product rule with $u = x^2 + 2x$ and $v = e^x$.

$$\frac{d^2 y}{dx^2} = u\frac{dv}{dx} + v\frac{du}{dx} = (x^2 + 2x)e^x + e^x(2x + 2) = (x^2 + 4x + 2)e^x$$

When $x = 0$, $\dfrac{d^2 y}{dx^2} = 2$. This is greater than 0, so $(0, 0)$ is a minimum.

When $x = -2$, $\dfrac{d^2 y}{dx^2} = (4 - 8 + 2)e^{-2}$. This is less than 0, so $(-2, 4e^{-2})$ is a maximum.

Exercise C (answers p 194)

1 Differentiate these with respect to x.

(a) $x^5 e^x$ (b) $x^{\frac{1}{2}} e^x$ (c) $x^2 \ln x$ (d) $3x e^x$

(e) $e^x \ln x$ (f) $e^x \sqrt[3]{x}$ (g) $x^{-2} \ln x$ (h) $x^{-2} e^x$

2 Find $f'(x)$ given that

(a) $f(x) = e^x(2x + 3)$ (b) $f(x) = (4x - 1)\ln x$ (c) $f(x) = e^x(x^2 + 1)$

(d) $f(x) = (\sqrt{x} - 1)\ln x$ (e) $f(x) = e^x\left(1 - \dfrac{1}{x}\right)$ (f) $f(x) = e^x(x + \ln x)$

3 The expression $x^2(x + 1)e^x$ can be rewritten as either $(x^3 + x^2)e^x$ or $x^3 e^x + x^2 e^x$.

(a) Use the product rule to differentiate $(x^3 + x^2)e^x$ with respect to x.

(b) Use the product rule to differentiate each of $x^3 e^x$ and $x^2 e^x$ and hence write down the derivative of $x^3 e^x + x^2 e^x$.
Check that the result agrees with (a).

(c) Find the derivative of $x^2(x + 1)\ln x$.

4 (a) Find $\dfrac{dy}{dx}$ given that $y = xe^x$.

(b) Show that the graph of $y = xe^x$ has a stationary point at $x = -1$.

(c) Find the y-coordinate of this stationary point.

(d) By finding $\dfrac{d^2y}{dx^2}$, determine the type of this stationary point.

5 (a) Given that $f(x) = e^x(x^2 - 7x + 13)$, find $f'(x)$.

(b) Express $f'(x)$ in the form $e^x(x - a)(x - b)$.

(c) Find the values of x for which $f(x)$ has a stationary value.

(d) Determine the type of each stationary value.

6 You are given that $y = x\ln x \ (x > 0)$.

(a) Find the gradient of the graph of $y = x\ln x$ at the point where $x = e$.

(b) Show that the point on the graph of $y = x\ln x$ where $x = \dfrac{1}{e}$ is a stationary point.

(c) Show that this stationary point is a minimum.

7 (a) Given that $y = x^2\ln x \ (x > 0)$, show that $\dfrac{dy}{dx} = x(1 + 2\ln x)$.

(b) Hence show that the graph of $y = x^2\ln x$ has a stationary point at $x = e^{-0.5}$.

(c) Determine the type of this stationary point.

D Differentiating a quotient (answers p 195)

In the previous section we looked at $\dfrac{dy}{dx}$ where y is the product uv of two functions.

In this section we consider the case where y is the quotient $\dfrac{u}{v}$ of two functions.

The relationship $y = \dfrac{u}{v}$ can be rewritten in product form as $u = yv$.

We can use the product rule to differentiate with respect to x: $\qquad \dfrac{du}{dx} = y\dfrac{dv}{dx} + v\dfrac{dy}{dx}$

Rearrange this equation to make $\dfrac{dy}{dx}$ the subject: $\qquad \dfrac{dy}{dx} = \dfrac{\dfrac{du}{dx} - y\dfrac{dv}{dx}}{v}$

Now use the fact that $y = \dfrac{u}{v}$ to replace y: $\qquad = \dfrac{\dfrac{du}{dx} - \dfrac{u}{v}\left(\dfrac{dv}{dx}\right)}{v}$

Finally multiply top and bottom by v: $\qquad = \dfrac{v\dfrac{du}{dx} - u\dfrac{dv}{dx}}{v^2}$

> **K**
>
> If $y = \dfrac{u}{v}$, then $\dfrac{dy}{dx} = \dfrac{v\dfrac{du}{dx} - u\dfrac{dv}{dx}}{v^2}$. This is called the **quotient rule**.
>
> It can also be written using function notation:
>
> The derivative of $\dfrac{f(x)}{g(x)}$ is $\dfrac{f'(x)g(x) - f(x)g'(x)}{[g(x)]^2}$.

D1 Use the quotient rule to differentiate each of these.

(a) $y = \dfrac{x^2}{x+1}$ (b) $y = \dfrac{x}{x^3 - 1}$ (c) $y = \dfrac{\ln x}{x}$ (d) $y = \dfrac{e^x}{x-1}$

Example 5

Given that $y = \dfrac{1+x^2}{1+e^x}$, find $\dfrac{dy}{dx}$.

Solution

Use the quotient rule with $u = 1 + x^2$ and $v = 1 + e^x$.

$$\dfrac{du}{dx} = 2x,\ \dfrac{dv}{dx} = e^x,\ \text{ so } \dfrac{dy}{dx} = \dfrac{v\dfrac{du}{dx} - u\dfrac{dv}{dx}}{v^2} = \dfrac{(1+e^x)2x - (1+x^2)e^x}{(1+e^x)^2}$$

This can be tidied up by collecting, in the numerator, the terms in e^x:

$$\dfrac{dy}{dx} = \dfrac{2x + e^x(-x^2 + 2x - 1)}{(1+e^x)^2}$$

Exercise D (answers p 195)

1 Differentiate these with respect to x.

(a) $\dfrac{x^2}{2x+3}$ (b) $\dfrac{x^3}{x-1}$ (c) $\dfrac{x}{x^2+1}$ (d) $\dfrac{2x}{x^2+x+1}$

(e) $\dfrac{x-1}{x^2-x+1}$ (f) $\dfrac{x}{2x^3-1}$ (g) $\dfrac{x^2}{x^3+1}$ (h) $\dfrac{x^2-1}{x^3+5x+1}$

2 (a) Use the quotient rule to differentiate $\dfrac{e^x}{x^2}$ with respect to x.

(b) The expression $\dfrac{e^x}{x^2}$ can also be written as a product $x^{-2}e^x$.

Use the product rule to differentiate $x^{-2}e^x$ and check that the result agrees with that of (a).

3 Find $f'(x)$ given that

(a) $f(x) = \dfrac{e^x}{x+1}$ (b) $f(x) = \dfrac{e^x}{x^2+1}$ (c) $f(x) = \dfrac{x^2}{e^x+1}$

(d) $f(x) = \dfrac{\ln x}{x-1}$ (e) $f(x) = \dfrac{\ln x}{x^2+1}$ (f) $f(x) = \dfrac{e^x}{\ln x}$

4 (a) Given that $y = \dfrac{x}{x+1}$, show that $\dfrac{dy}{dx} = \dfrac{1}{(x+1)^2}$.

(b) Hence show that there are no stationary points on the graph of $y = \dfrac{x}{x+1}$.

5 (a) Given that $f(x) = \dfrac{x+1}{(x^2+3)}$, show that $f'(x) = \dfrac{-(x-1)(x+3)}{(x^2+3)^2}$.

(b) Hence find the values of x for which $f(x)$ is stationary.

6 Given that $f(x) = \dfrac{x}{(x^2+4)}$, find the values of x for which $f'(x) = 0$.

7 Find the coordinates of the stationary point on the graph of $y = \dfrac{e^x}{x-2}$.

8 (a) Given that $y = \dfrac{\ln x}{x}$ $(x > 0)$, find (i) $\dfrac{dy}{dx}$ (ii) $\dfrac{d^2y}{dx^2}$

(b) Find the coordinates of the stationary point on the graph of $y = \dfrac{\ln x}{x}$.

(c) Determine the type of this stationary point.

9 Show that the graph of $y = \dfrac{x^2}{\ln x}$ $(x > 0)$ has just one stationary point whose coordinates are $(e^{0.5}, 2e)$.

E Differentiating sin x, cos x and tan x (answers p 196)

Here is the graph of $y = \sin x$, where x is in radians.

The diagram below shows a close-up of the graph as it passes through the origin O.

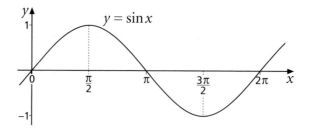

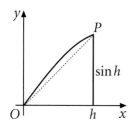

The coordinates of the point P on the graph are $(h, \sin h)$, where h is small.

The gradient of OP is $\dfrac{\sin h}{h}$.

E1 Find the value of $\dfrac{\sin h}{h}$ when $h = 0.1, 0.01$ and 0.001.

(Make sure your calculator is set to work in radians.)

What conclusion do you draw about the gradient of $y = \sin x$ at $x = 0$?

D

E2 (a) What is the gradient of $y = \sin x$ at each of these points?

$$x = \frac{\pi}{2} \qquad x = \pi \qquad x = \frac{3\pi}{2} \qquad x = 2\pi$$

(b) From the graph above, you can see that between $x = 0$ and $x = \dfrac{\pi}{2}$ the gradient of $y = \sin x$ decreases from 1 to 0, slowly at first but then more rapidly.

Sketch the graph of the gradient $\dfrac{dy}{dx}$ between $x = 0$ and $x = \dfrac{\pi}{2}$ and extend the sketch graph up to $x = 2\pi$.

What do you think its equation is?

E3 Here is the graph of $y = \cos x$.

Sketch the graph of $\dfrac{dy}{dx}$.

What do you think its equation is?

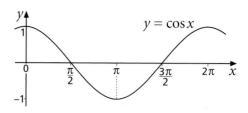

It can be proved, by methods beyond the scope of this book, that

The derivative of $\sin x$ is $\cos x$.

The derivative of $\cos x$ is $-\sin x$.

An important reason for using radians is that they give simple results like these.

E4 Write down the derivative of each of these.

(a) $\sin x + 3x + x^2$ (b) $\sqrt{x} - \cos x$ (c) $3e^x + 4\sin x$ (d) $5\ln x - 2\cos x$

To differentiate $\tan x$, we use the fact that $\tan x = \dfrac{\sin x}{\cos x}$.

E5 (a) Given that $y = \tan x = \dfrac{\sin x}{\cos x}$, use the quotient rule to show that $\dfrac{dy}{dx} = \dfrac{\cos^2 x + \sin^2 x}{\cos^2 x}$.

(b) Explain why this expression for $\dfrac{dy}{dx}$ is equivalent to $\sec^2 x$.

Ⓚ The derivative of $\tan x$ is $\sec^2 x$.

E6 What happens to $\sec^2 x$ as x gets closer and closer to $\dfrac{\pi}{2}$?
How is this related to the behaviour of the graph of $y = \tan x$?

Example 6

Find $f'(x)$ given that (a) $f(x) = e^x \sin x$ (b) $f(x) = \dfrac{\tan x}{x^2 + 1}$

Solution

(a) Use the product rule. $f'(x) = e^x \cos x + (\sin x)e^x = e^x(\cos x + \sin x)$

(b) Use the quotient rule. $f'(x) = \dfrac{(x^2 + 1)\sec^2 x - (\tan x)2x}{(x^2 + 1)^2} = \dfrac{(x^2 + 1)\sec^2 x - 2x\tan x}{(x^2 + 1)^2}$

Exercise E (answers p 196)

1 Differentiate these with respect to x.

(a) $\sin x + \cos x$ (b) $\sin x - x^2$ (c) $3\sin x - 2\cos x$ (d) $x + 2\tan x$

2 Differentiate these with respect to x.

(a) $3x\cos x$ (b) $\dfrac{\cos x}{x}$ (c) $\dfrac{\sin x}{x^2}$ (d) $e^x \tan x$

3 Given that $f(x) = \dfrac{\sin x}{x}$, show that $f'(x) = \dfrac{\cos x}{x} - \dfrac{\sin x}{x^2}$.

4 (a) Find $f'(x)$ given that $f(x) = 3\sin x + 4\cos x$.

(b) Show that if $f'(x) = 0$, then $\tan x = 0.75$ and find the values of x in the interval $0 \le x \le 2\pi$ for which $f'(x) = 0$.

5 Given that $f(x) = \dfrac{\cos x}{x^2}$, show that $f'\left(\dfrac{\pi}{2}\right) = -\dfrac{4}{\pi^2}$.

6 Given that $f(x) = \sin x - \tan x$, find the values of x in the interval $0 \le x \le 2\pi$ for which $f'(x) = 0$.

F Differentiating a function of a function (answers p 196)

$y = \sin(x^2 - 3)$ is an example of a function of a function.

If we let $u = x^2 - 3$, then $y = \sin u$. So y is a function of u, which is itself a function of x.

The general case, where u and y are any functions, can be pictured like this:

Input $x \to u \to y$

Suppose x is increased by a small quantity δx and that, as a result, u is increased by δu and y by δy.

The following relationship can be seen to be true by 'cancelling out' δu: $\dfrac{\delta y}{\delta x} = \dfrac{\delta y}{\delta u} \times \dfrac{\delta u}{\delta x}$

As δx gets smaller and smaller, $\dfrac{\delta y}{\delta x}$ gets closer and closer to $\dfrac{dy}{dx}$, $\dfrac{\delta y}{\delta u}$ to $\dfrac{dy}{du}$ and $\dfrac{\delta u}{\delta x}$ to $\dfrac{du}{dx}$.

So we get the following relationship:

$$\frac{dy}{dx} = \frac{dy}{du} \times \frac{du}{dx}$$

This relationship is called the **chain rule**.

F1 In the example at the top of this page, $y = \sin u$ where $u = x^2 - 3$.

(a) Find $\dfrac{dy}{du}$.

(b) Find $\dfrac{du}{dx}$.

(c) Use the chain rule to show that $\dfrac{dy}{dx} = 2x\cos(x^2 - 3)$.

F2 The function $y = \sin^2 x$ can be written as $y = u^2$, where $u = \sin x$.

Find $\dfrac{dy}{dx}$.

F3 The function $y = \sqrt{x^2 + 1}$ can be written as $y = u^{\frac{1}{2}}$, where $u = x^2 + 1$.

Show that $\dfrac{dy}{dx} = \dfrac{x}{\sqrt{x^2 + 1}}$.

Example 7

Find $\dfrac{dy}{dx}$ given that $y = \ln(1 + x^2)$.

Solution

Let $u = 1 + x^2$, so that $y = \ln u$. So $\dfrac{du}{dx} = 2x$ and $\dfrac{dy}{du} = \dfrac{1}{u}$.

$\dfrac{dy}{dx} = \dfrac{dy}{du} \times \dfrac{du}{dx} = \dfrac{1}{u} \times 2x = \dfrac{2x}{1 + x^2}$

The simplest type of 'function of a function' is where the 'intermediate' function u is a linear function, such as $u = 2x + 3$.

For this type, $\dfrac{du}{dx}$ is just a number.

After some practice you should find that you can do some of the working in your head, as shown in the next example.

Example 8

Find $\dfrac{dy}{dx}$ given that $y = \sin(3x + 1)$.

Solution

Full working:

Let $u = 3x + 1$, so $y = \sin u$.

$$\frac{dy}{dx} = \frac{dy}{du} \times \frac{du}{dx} = \cos u \times 3 = 3\cos(3x + 1)$$

Working in head:

Derivative of $\sin$ *is* $\cos$.

Derivative of $3x + 1$ *is* 3.

So $\dfrac{dy}{dx} = 3\cos(3x + 1)$

F4 Find the derivative of each of these.

(a) $\cos 2x$

(b) e^{-3x}

(c) $\ln(6x + 1)$

(d) $\sqrt{5 - 2x}$

(e) $\dfrac{1}{3 + 4x}$

(f) $\sin(5x - 2)$

(g) $(3x - 2)^{\frac{2}{3}}$

(h) $\dfrac{1}{\sqrt{1 + 4x}}$

It is worth learning, as a commonly occurring special case, that

K The derivative of $f(ax)$ is $af'(ax)$.

For example, the derivative of $\sin 4x$ is $4\cos 4x$, the derivative of e^{-5x} is $-5e^{-5x}$, and so on.

The chain rule can be extended.

For example, the function $y = \sqrt{\sin(x^2)}$ can be broken down as $y = \sqrt{u}$, $u = \sin v$, $v = x^2$.

The chain rule is then $\dfrac{dy}{dx} = \dfrac{dy}{du} \times \dfrac{du}{dv} \times \dfrac{dv}{dx}$.

F5 Use the extended chain rule to differentiate $y = \sqrt{\sin(x^2)}$.

The chain rule (with one 'intermediate' function) can be expressed in function notation as: The derivative of $f(g(x))$ is $f'(g(x))g'(x)$.

Exercise F (answers p 197)

1 Differentiate these with respect to x.

(a) $\sin(3x - 2)$

(b) $\cos(5x + 1)$

(c) $\sin(3 - 2x)$

(d) $\cos\left(\frac{1}{2}x\right)$

(e) $\ln(3x + 1)$

(f) $\sqrt{5x - 1}$

(g) $(2x + 1)^7$

(h) $\tan(4x)$

(i) $(2x + 1)^{\frac{1}{3}}$

(j) $\tan(3x + 2)$

(k) $e^{5x - 2}$

(l) $\cos\left(1 - \frac{1}{2}x\right)$

2 If $y = \cos^2 x$ then $y = u^2$ where $u = \cos x$.

Find $\dfrac{dy}{du}$ and $\dfrac{du}{dx}$ and use the chain rule to find $\dfrac{dy}{dx}$.

3 If $y = \ln(\sin x)$ then $y = \ln u$ where $u = \sin x$.

Find $\dfrac{dy}{du}$ and $\dfrac{du}{dx}$ and use the chain rule to find $\dfrac{dy}{dx}$.

4 Differentiate $y = e^{\sin x}$ by letting $u = \sin x$ and using the chain rule.

5 Find $\dfrac{dy}{dx}$ given that

(a) $y = \sin(e^x)$ **(b)** $y = (\ln x)^2$ **(c)** $y = \tan^2 x$ **(d)** $y = (x^2 + 1)^5$

6 Find $f'(x)$ given that

(a) $f(x) = e^{\tan x}$ **(b)** $f(x) = \sqrt{\ln x}$ **(c)** $f(x) = \dfrac{1}{\ln x}$ **(d)** $f(x) = \sin(\sqrt{x})$

7 Find $\dfrac{dy}{dx}$ given that

(a) $y = \ln(x^2 + 1)$ **(b)** $y = (\ln x)^2$ **(c)** $y = e^{x^2}$ **(d)** $\tan(e^x)$

(e) $y = \sin(1 - x^2)$ **(f)** $y = \sqrt{1 + x^2}$ **(g)** $y = \dfrac{1}{1 + x^2}$ **(h)** $y = \dfrac{1}{(1 + x^2)^3}$

8 (a) Given that $y = \dfrac{1}{\sqrt{1 - x^2}}$, show that $\dfrac{dy}{dx} = \dfrac{x}{\left(1 - x^2\right)^{\frac{3}{2}}}$.

(b) Find the gradient of the graph of y at the point where $x = 0.6$.

G Further trigonometrical functions (answers p 197)

G1 If $y = \operatorname{cosec} x = \dfrac{1}{\sin x}$, then $y = u^{-1}$ where $u = \sin x$.

(a) Find $\dfrac{dy}{du}$ and $\dfrac{du}{dx}$ and use the chain rule to show that $\dfrac{dy}{dx} = -\dfrac{\cos x}{\sin^2 x}$.

(b) Show that this result can also be written as $-\operatorname{cosec} x \cot x$.

G2 If $y = \sec x = \dfrac{1}{\cos x}$, then $y = u^{-1}$ where $u = \cos x$.

Show that $\dfrac{dy}{dx} = \dfrac{\sin x}{\cos^2 x} = \sec x \tan x$.

G3 If $y = \cot x = \dfrac{1}{\tan x}$, then $y = u^{-1}$ where $u = \tan x$.

Show that $\dfrac{dy}{dx} = \dfrac{-\sec^2 x}{\tan^2 x} = -\operatorname{cosec}^2 x$.

G4 Find the derivative of $\cot x$ in a different way, by using the fact that $\cot x = \dfrac{\cos x}{\sin x}$ and the quotient rule. Check that the result agrees with that in G3.

In section B above, it was shown that $\dfrac{dy}{dx} = \dfrac{1}{\dfrac{dx}{dy}}$.

This relationship is sometimes needed in order to find $\dfrac{dy}{dx}$.

G5 Given that $x = \sin y$,

(a) find $\dfrac{dx}{dy}$

(b) use the fact that $\cos^2 y + \sin^2 y = 1$ to show that $\dfrac{dy}{dx} = \dfrac{1}{\sqrt{1-x^2}}$

Example 9

Given that $y = \sec^2 3x$, find $\dfrac{dy}{dx}$.

Solution

Use the chain rule.   Let $u = \sec 3x$, so $y = u^2$.

Then $\dfrac{du}{dx} = 3\sec 3x \tan 3x$ and $\dfrac{dy}{du} = 2u$.

$\dfrac{dy}{dx} = \dfrac{dy}{du} \times \dfrac{du}{dx}$

$= 2u \times 3\sec 3x \tan 3x$

$= 2\sec 3x \times 3\sec 3x \tan 3x$

$= 6\sec^2 3x \tan 3x$

Example 10

Given that $x = \sin 2y$, show that $\dfrac{dy}{dx} = \dfrac{1}{2\sqrt{1-x^2}}$.

Solution

x is given as a function of y, so first find $\dfrac{dx}{dy}$. $\dfrac{dx}{dy} = 2\cos 2y$.

Use the relationship between $\dfrac{dy}{dx}$ and $\dfrac{dx}{dy}$. $\dfrac{dy}{dx} = \dfrac{1}{\dfrac{dx}{dy}} = \dfrac{1}{2\cos 2y}$

To get the result in terms of x, use the fact that $\cos^2 2y + \sin^2 2y = 1$.

$\cos 2y = \sqrt{1 - \sin^2 2y} = \sqrt{1 - x^2}$, so $\dfrac{dy}{dx} = \dfrac{1}{2\sqrt{1-x^2}}$.

Exercise G (answers p 197)

1 Find the derivative of

(a) $\operatorname{cosec} 2x$ (b) $\cot 4x$ (c) $\sec(3x - 1)$ (d) $\operatorname{cosec}(x^2 + 1)$

(e) $\cot\left(\dfrac{1}{x}\right)$ (f) $\operatorname{cosec}(\sqrt{x})$ (g) $\sec(1 - x^3)$ (h) $\cot(x^2 + x)$

2 Show that the graph of $y = \sec x + \operatorname{cosec} x$ has a stationary point where $x = \dfrac{\pi}{4}$.

3 Show that the graph of $y = x - \cot x$ has no stationary point.

4 Given that $y = \sec^2 x$, use the chain rule to show that $\dfrac{dy}{dx} = 2\sec^2 x \tan x$.

5 Find the derivative of

(a) $\operatorname{cosec}^2 x$ (b) $\cot^2 x$ (c) $\sec^3 x$ (d) $\operatorname{cosec}^2 3x$

(e) $\sqrt{(\cot x)}$ (f) $e^{\sec x}$ (g) $\sec(e^x)$ (h) $\cot\left(2\sqrt{x}\right)$

6 Given that $x = \cos y$,

(a) find $\dfrac{dx}{dy}$ (b) show that $\dfrac{dy}{dx} = -\dfrac{1}{\sqrt{1 - x^2}}$

7 Given that $x = \tan y$,

(a) find $\dfrac{dx}{dy}$ (b) show that $\dfrac{dy}{dx} = \dfrac{1}{1 + x^2}$

8 Given that $x = \sec y$,

(a) find $\dfrac{dx}{dy}$ (b) show that $\dfrac{dy}{dx} = \dfrac{1}{x\sqrt{x^2 - 1}}$

9 Given that $x = \cos 3y$, show that $\dfrac{dy}{dx} = -\dfrac{1}{3\sqrt{1 - x^2}}$

*10 Given that $y = \sec x + \tan x$, find $\dfrac{dy}{dx}$ and show that it can be written in the form $\dfrac{1}{1 - \sin x}$.

H Selecting methods (answers p 198)

When you differentiate a function you need to decide which rule to apply:

     Product rule   Quotient rule   Chain rule

The special case of the chain rule, where you have a function of $(ax + b)$, is very common and is worth learning separately:

Ⓚ   Derivative of $f(ax + b) = af'(ax + b)$

H1 Which rule would you use if you had to differentiate each of these?

(a) $\tan(2x - 1)$ (b) $\ln(\tan x)$ (c) $\dfrac{\ln x}{\tan x}$ (d) $\ln x \tan x$

You may need to use more than one of the rules.

For example, if $y = e^x \sin 3x$, then you can use the product rule with $u = e^x$ and $v = \sin 3x$. To use the product rule you need to find $\dfrac{du}{dx}$ and $\dfrac{dv}{dx}$.

Now $\sin 3x$ is itself a function of a function. Its derivative is $3 \cos 3x$.

So we have $\dfrac{dy}{dx} = u \dfrac{dv}{dx} + v \dfrac{du}{dx} = e^x(3 \cos 3x) + (\sin 3x)e^x = e^x(3 \cos 3x + \sin 3x)$.

H2 Given that $y = e^{2x} \sin x$, let $u = e^{2x}$ and $v = \sin x$.

    (a) Find $\dfrac{du}{dx}$ and $\dfrac{dv}{dx}$.   **(b)** Find $\dfrac{dy}{dx}$.

H3 Given that $y = \dfrac{e^x}{\sin 4x}$, let $u = e^x$ and $v = \sin 4x$.

    Find $\dfrac{du}{dx}$ and $\dfrac{dv}{dx}$ and hence find $\dfrac{dy}{dx}$.

H4 Given that $y = e^{2x} \cos 3x$, let $u = e^{2x}$ and $v = \cos 3x$.

    Find $\dfrac{du}{dx}$ and $\dfrac{dv}{dx}$ and hence find $\dfrac{dy}{dx}$.

For a function like $x^2 e^x + 3 \ln x$ you can use the product rule for the first term, $x^2 e^x$, and then add the derivative of the second term, $3 \ln x$.

H5 Differentiate $x^2 e^x + 3 \ln x$.

Students sometimes have difficulty deciding when to use the product rule and when to use the chain rule.

H6 Differentiate   **(a)** $\sin (x^2 + 1)$ **(b)** $(x^2 + 1) \sin x$

Example 11

Find the gradient of the graph of $y = x^2 e^{-3x}$ at the point where $x = 1$.

Solution

y is the product of the two functions x^2 and e^{-3x}.

Let $u = x^2$ and $v = e^{-3x}$. Then $\dfrac{du}{dx} = 2x$ and $\dfrac{dv}{dx} = -3e^{-3x}$.

So $\dfrac{dy}{dx} = u \dfrac{dv}{dx} + v \dfrac{du}{dx} = x^2(-3e^{-3x}) + (e^{-3x})2x = e^{-3x}(-3x^2 + 2x)$.

When $x = 1$, $\dfrac{dy}{dx} = e^{-3}(-3 + 2) = -e^{-3}$.

Exercise H (answers p 198)

1 Find $\dfrac{dy}{dx}$ given that

(a) $y = x^2(e^x + 1)$ (b) $y = \ln(x^2 - 2)$ (c) $y = \ln(\cos x)$ (d) $y = \ln x \sin x$

(e) $y = \sin(\ln x)$ (f) $y = \dfrac{e^x}{5x + 1}$ (g) $y = \dfrac{e^{-x}}{2x + 1}$ (h) $y = \dfrac{e^{-2x}}{1 - x}$

2 Find $f'(x)$ given that

(a) $f(x) = e^x \sin 3x$ (b) $f(x) = e^{-x} \cos 3x$ (c) $f(x) = e^{-2x} \tan x$

(d) $f(x) = \sin 3x \cos 2x$ (e) $f(x) = \tan(2 - 3x)$ (f) $f(x) = \dfrac{\ln x}{1 - 2x}$

3 Show that the gradient of the curve $y = \dfrac{\sin 2x}{x}$ at the point where $x = \dfrac{\pi}{2}$ is $-\dfrac{4}{\pi}$.

4 Find the gradient of each of these curves at the point where $x = 0$.

(a) $y = xe^{2x} + \sin x$ (b) $y = (2x + 1)e^{-x}$ (c) $y = \tan x - \sin 2x$

Key points

- The derivative of e^x is e^x.   (p 114)

- The derivative of $\ln x$ is $\dfrac{1}{x}$.   (p 115)

- $\dfrac{dy}{dx} = \dfrac{\frac{1}{dx}}{dy}$ (p 115)

- If $y = uv$, then $\dfrac{dy}{dx} = u\dfrac{dv}{dx} + v\dfrac{du}{dx}$ (product rule)   (p 117)

- If $y = \dfrac{u}{v}$, then $\dfrac{dy}{dx} = \dfrac{v\frac{du}{dx} - u\frac{dv}{dx}}{v^2}$ (quotient rule)   (p 120)

- The derivative of $\sin x$ is $\cos x$.
  The derivative of $\cos x$ is $-\sin x$.   (p 122)

- The derivative of $\tan x$ is $\sec^2 x$.   (p 123)

- $\dfrac{dy}{dx} = \dfrac{dy}{du} \times \dfrac{du}{dx}$ (chain rule)   (p 124)

- The derivative of $f(ax)$ is $af'(ax)$.
  The derivative of $f(ax + b)$ is $af'(ax + b)$.   (p 125)

- The derivative of $\operatorname{cosec} x$ is $-\operatorname{cosec} x \cot x$.
  The derivative of $\sec x$ is $\sec x \tan x$.
  The derivative of $\cot x$ is $-\operatorname{cosec}^2 x$.   (p 127)

Mixed questions (answers p 198)

1 Find the gradient of the curve $y = \dfrac{\tan x}{x+2}$ at the point where $x = 0$.

2 Differentiate these with respect to x.

(a) e^{3x-2} (b) $\ln(3x-2)$ (c) $(3x-2)\ln x$ (d) $(3x-1)^4$

3 Given that $y = 5\ln(1 + 0.1x)$, find the value of x for which $\dfrac{dy}{dx} = 0.4$.

4 Show that the graphs of $y = x^2 - 5x$, $y = \dfrac{1}{x^2-5x}$ and $y = \ln(x^2 - 5x)$

all have a stationary point at the same value of x and find this value of x.

5 Given that $x = \tan 3y$, prove that $\dfrac{dy}{dx} = \dfrac{1}{3(1+x^2)}$.

6 Given that $y = e^x\cos 3x$, show that $\dfrac{dy}{dx}$ can be expressed in the form $Re^x\cos(3x + \alpha)$.

Find, to three significant figures, the values of R and α, where $0 < \alpha < \dfrac{\pi}{2}$.

Test yourself (answers p 199)

1 As a substance cools, its temperature, $T\,°C$, is related to the time (t minutes) for which it has been cooling. The relationship is given by the equation

$$T = 20 + 60e^{-0.1t}, \ t \geq 0$$

(a) Find the value of T when the substance started to cool.

(b) Explain why the temperature of the substance is always above $20\,°C$.

(c) Sketch the graph of T against t.

(d) Find the value, to two significant figures, of t at the instant $T = 60$.

(e) Find $\dfrac{dT}{dt}$.

(f) Hence find the value of T at which the temperature is decreasing at a rate of $1.8\,°C$ per minute.

Edexcel

2 Differentiate these with respect to x.

(a) $\sin(3x-2)$ (b) $e^{5\sqrt{x}}$ (c) $\tan(4-x)$ (d) $\ln(7-x^2)$

3 Differentiate these with respect to x.

(a) $e^{-x}\sin x$ (b) $\dfrac{e^x}{\cos x}$ (c) $(x^2-7)^9$ (d) $\dfrac{x^2}{\tan x}$

4 Find the gradient of the graph of $y = x\sin 2x$ at the point where $x = \dfrac{\pi}{4}$.

5 Given that $x = \sin\frac{1}{2}y$, prove that $\dfrac{dy}{dx} = \dfrac{2}{\sqrt{1-x^2}}$.

9 Numerical methods

In this chapter you will learn how to
- confirm that an equation has a solution between two values
- find an approximate solution to an equation using an iterative formula

A Locating roots (answers p 199)

A1 Given the function $f(x) = 3 + x - x^2$, check that $f(2) = 1$ and $f(3) = -3$.

These values suggest that the graph of $y = 3 + x - x^2$ crosses the x-axis between $x = 2$ and $x = 3$. The value of x where it does so is a solution of the equation $3 + x - x^2 = 0$.
So the fact that $f(x)$ changes from positive to negative between $x = 2$ and $x = 3$ tells us that there is a solution of the equation in that interval.

As this is a quadratic graph it was safe to assume that it continues without a break between $f(2)$ and $f(3)$.
But if it had been a function with a break as shown here, we could not deduce that it crosses the x-axis between $x = 2$ and $x = 3$. So we could not deduce that a solution of $f(x) = 0$ lies between those values.

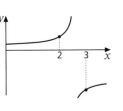

The function in the first diagram is **continuous**. The function in the second diagram is not continuous (it has a **discontinuity**) between $x = 2$ and $x = 3$.

Many functions (such as polynomials, e^x, $\sin x$) are continuous, but functions such as $\dfrac{1}{x-1}$ and $\tan x$ have discontinuities.

K For an equation of the form $f(x) = 0$, if $f(x_1)$ and $f(x_2)$ have opposite signs and $f(x)$ is continuous between x_1 and x_2, then a root (solution) of the equation lies between x_1 and x_2.

There could be more than one root between the two values of x, as this diagram shows.

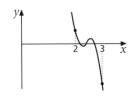

Example 1

Show that the equation $\dfrac{x^3}{5} = x(x - 1)$ has a root between 3 and 4.

Solution

Rearrange the equation in the form $f(x) = 0$. $x^3 - 5x^2 + 5x = 0$

$x^3 - 5x^2 + 5x$ is a polynomial so is continuous.

When $x = 3$, the polynomial takes the value $3^3 - 5 \times 3^2 + 5 \times 3 = -3$, and when $x = 4$, it takes the value $4^3 - 5 \times 4^2 + 5 \times 4 = 4$.

The sign changes so the equation has a root between $x = 3$ and $x = 4$.

'The interval $(1, 2)$' means all values between 1 and 2, but not 1 and 2 themselves, while 'the interval $[1, 2]$' means all values from 1 to 2 inclusive.

Example 2

The graphs of $y = x$ and $y = \sqrt{x+1}$ are shown.

(a) Show that the x-coordinate of the point of intersection P satisfies the equation $x^2 - x - 1 = 0$.

(b) Show that the equation in (a) has a root in the interval $(1, 2)$.

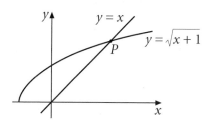

Solution

(a) *At P the y-value for each graph is the same.* $\qquad x = \sqrt{x+1}$

$\qquad$ *Square both sides.* $\qquad\qquad\qquad\qquad x^2 = x + 1$

$$\Rightarrow x^2 - x - 1 = 0$$

(b) $\qquad\qquad\qquad\qquad$ When $x = 1$, $x^2 - x - 1 = 1^2 - 1 - 1 = -1$

$\qquad\qquad\qquad\qquad$ and when $x = 2$, $x^2 - x - 1 = 2^2 - 2 - 1 = 1$

$\qquad\qquad\qquad\qquad$ The sign changes and the function $x^2 - x - 1$ is continuous, so there is a root in the interval $(1, 2)$.

Example 3

α is the positive root of the equation $x^3 - \dfrac{1}{x} - 1 = 0$. Show that, to 3 d.p., $\alpha = 1.221$.

Solution

For $\alpha = 1.221$ to be the root correct to 3 d.p., α must lie between 1.2205 and 1.2215.

When $x = 1.2205$, $x^3 - \dfrac{1}{x} - 1 = -0.00125\ldots$

When $x = 1.2215$, $x^3 - \dfrac{1}{x} - 1 = 0.00388\ldots$

The sign changes so α lies between 1.2205 and 1.2215. Hence, to 3 d.p., $\alpha = 1.221$.

Exercise A (answers p 199)

Remember to set your calculator to radians.

1 Show that the equation $\ln x - 1 = 0$ has a root between 2 and 3.

2 Show that the equation $e^x - 3x^2 = 0$ has roots in the intervals $(-1, 0)$, $(0, 1)$ and $(3, 4)$.

3 (a) Show that the equation $\sin x - \dfrac{1}{x} = 0$ has roots in the interval $(-3, -2)$ and $(2, 3)$.

$\qquad$ (b) Does it have a root in the interval $(-1, 1)$? Give a reason for your answer.

4 By rewriting the equation $\cos x = \ln x$ in the form $f(x) = 0$ show that it has at least one root between 1 and 2.

5 By rewriting the equation $\sqrt{x} = x^2 - 2$ in a suitable form, show that it has at least one root in the interval $(1.8, 1.9)$.

6 The diagram shows the graphs of $y = x^3 - 10$ and $y = \dfrac{1}{x}$.

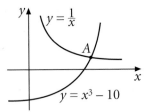

(a) Show that the x-coordinate of the point A satisfies the equation $x^4 - 10x - 1 = 0$.

(b) Show that the x-coordinate of A lies between 2 and 3.

7 (a) On the same axes, sketch the graphs of $y = 2\cos x$ and $y = 1 - \dfrac{x^2}{4}$.

(b) From your sketch, state how many roots the equation $2\cos x = 1 - \dfrac{x^2}{4}$ has.

(c) Show that the equation $2\cos x = 1 - \dfrac{x^2}{4}$ has a root between -1.3 and -1.2.

8 You are told that, for a given continuous function f(x), f(2) and f(3) are both positive. Does it follow that f(x) = 0 has no root between 2 and 3? Give a reason.

9 Show that the equation $x^4 - 2x^3 = 4$ has a root α between $-1.089\,85$ and $-1.089\,95$. Hence write the value of α to five significant figures.

10 Show that the equation $e^{-x} = x^3$ has a root which, correct to three significant figures, is 0.773.

11 α is the only root of the equation $e^{2x} - \dfrac{1}{x^2} = 0$. By considering suitable values of x, show that $\alpha = 0.5671$ to four significant figures.

12 Show that the equation $\sqrt{x} + \cos x - 1 = 0$ has a root in the interval $(1.9965, 1.9975)$. On the basis of this information, what is the greatest degree of accuracy to which you can state the value of this root?

B Staircase and cobweb diagrams (answers p 200)

The equation $x^3 - 12x + 12 = 0$ can be written as $x^3 + 12 = 12x$ and hence as $\dfrac{x^3}{12} + 1 = x$.

This can be represented by two functions that we can graph:
$$y = \dfrac{x^3}{12} + 1 \qquad (1)$$
$$y = x \qquad (2)$$

The x-coordinate of a point of intersection of these two graphs is a root of the original equation $x^3 - 12x + 12 = 0$.

B1 (a) On the same pair of axes, sketch the two graphs or plot them on a graph plotter.

(b) How many points of intersection are there?

(c) Confirm that there is a point of intersection between $x = 1$ and $x = 2$.

On the next page we shall home in on this point of intersection, taking $x = 2$ as a starting value.

B2 Confirm the following steps on your calculator.
Keep each result in your calculator ready for the next step.

Substituting $x = 2$ into $y = \dfrac{x^3}{12} + 1$
gives $y = 1.66\ldots$

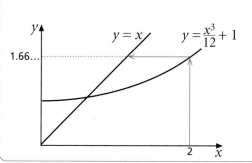

As the point marked ✄ is on $y = x$, its
coordinates are $(1.66\ldots, 1.66\ldots)$.
We shall use $1.66\ldots$ as our next value of x.

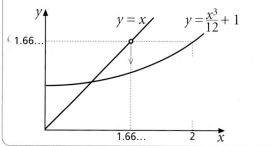

Substituting $x = 1.66\ldots$ into $y = \dfrac{x^3}{12} + 1$
gives $y = 1.38\ldots$

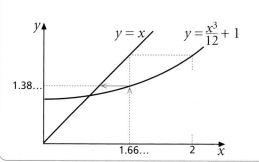

The point marked ✄ is $(1.38\ldots, 1.38\ldots)$.
We shall use $1.38\ldots$ as our next value of x.

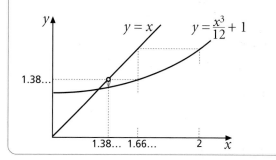

B3 Substitute $x = 1.38\ldots$ from your calculator into $y = \dfrac{x^3}{12} + 1$.

Repeat the process a few times, each time substituting the value obtained for y
back into the equation as the value of x.

You should find that the values converge to $1.1157\ldots$ This means that
the graphs (1) and (2) intersect at $(1.1157\ldots, 1.1157\ldots)$.

B4 Verify by substitution that $x = 1.1157$ is an approximate solution to
the original equation $x^3 - 12x + 12 = 0$.

The repeated substitution method can be represented by a 'staircase diagram'.
Each vertical arrow shows a value of x being input into the curved graph.
The subsequent horizontal arrow shows the value of y that is the output.
Because the points marked ✄ are on the graph $y = x$, each output
y-value becomes the next input x-value.

This is why we chose $y = x$ as one of the equations into which
we arranged the original equation, $x^3 - 12x + 12 = 0$.

Repeatedly taking the output from an equation and making it the new input can be described by a **recurrence relation** (or **iterative formula**) as used for sequences in Core 1. The recurrence relation you have just been using is

$$x_0 = 2, \quad x_{n+1} = \frac{x_n^3}{12} + 1$$

We used $x = 2$ as the starting value because we knew a point of intersection lay between $x = 1$ and $x = 2$. Now we will try using $x = 1$ as the starting value.

B5 Obtain the first few values from the recurrence relation $x_0 = 1, \quad x_{n+1} = \frac{x_n^3}{12} + 1$.
To what value do they converge?

You should find that starting with $x_0 = 1$ leads to the same limit as before, but here you are going up a staircase, not down.

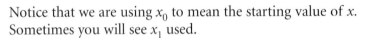

Notice that we are using x_0 to mean the starting value of x.
Sometimes you will see x_1 used.

With a recurrence relation it can take many steps for values to settle down to a required number of decimal places. A calculator with an ANS key helps. For example, for question B5, key $1 =$ (or 1 EXE) to enter the starting value. Without clearing, key ANS^3/12+1 then key $=$ (or EXE) as many times as required.

Alternatively you can use a spreadsheet, as here, or a similar facility on a graphic calculator.

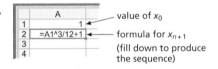

Using an iterative formula to solve an equation may not result in a staircase, as the following example shows.

B6 Show that there is a solution of $e^{-x} - x = 0$ between $x = 0.2$ and $x = 1$.

The equation $e^{-x} - x = 0$ can be rewritten as $x = e^{-x}$ and hence can be solved by finding where the graphs $y = x$ and $y = e^{-x}$ intersect.

B7 Starting with $x_0 = 0.2$, obtain the first few values from $x_{n+1} = e^{-x_n}$.
What happens? Do the values approach a limit?

The situation in B7 is shown in the diagram.
Again each vertical arrow shows an x-value being input into the curved graph, and the subsequent horizontal arrow shows the y-value that is the output.
Points marked ✗ on the graph $y = x$ again turn each output value into the next input.

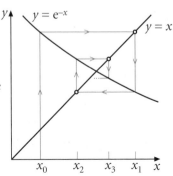

But the result is a 'cobweb', not a staircase, and although the values get closer to a limit, they are alternately greater and less than it, rather than to one side of it.

D **B8** **(a)** Trace each diagram and, starting from the x_0 marked, draw a staircase or cobweb as appropriate. Make sure your first vertical line ends on the **curved** graph.

(i)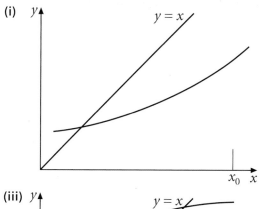

(ii)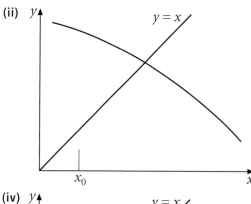

(iii)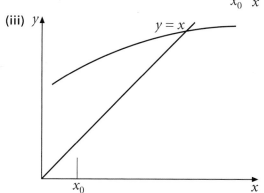

(iv)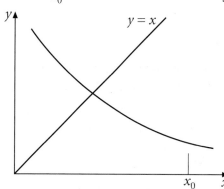

(b) What determines whether you get a staircase or a cobweb?

K When an iterative formula of the form $x_{n+1} = f(x_n)$ converges to a limit, the value of the limit is the x-coordinate of the point of intersection of the graphs of

$y = f(x)$ and $y = x$

The limit is therefore a solution of the equation $f(x) = x$.

A staircase or cobweb diagram based on the graphs of $y = f(x)$ and $y = x$ illustrates the convergence.

Using an iterative formula to solve an equation is an example of a **numerical method**. It is a useful approach when an equation cannot be solved entirely by algebra.

Example 4

Find, to 3 d.p., the limit which the recurrence relation $x_0 = 3$, $x_{n+1} = 4 - \ln x_n$ approaches.

Solution

If you have an ANS facility, key $3 = $ *(or 3 EXE) then* $4 - \ln \text{ANS}$ *then repeated* $=$.

$x_1 = 2.901...$, $x_2 = 2.934...$, $x_3 = 2.923...$; *after 7 iterations the values settle down to* 2.9262...

To 3 d.p. the limit is 2.926.

Example 5

(a) Using the iterative formula $x_{n+1} = \sqrt[3]{x_n + 1}$ with $x_0 = 1.5$, find x_1, x_2 and x_3 to 4 s.f.

(b) Given that the limit of of the sequence is L, show that L satisfies the equation $L^3 - L - 1 = 0$.

Solution

(a) *Write each value to 4 s.f., but keep the unrounded value in your calculator ready for the next stage.*
$$x_1 = 1.357, \quad x_2 = 1.331, \quad x_3 = 1.326$$

(b) At the limit, $L = x_n = x_{n+1}$

Substituting into the iterative formula, $\qquad L = \sqrt[3]{L+1}$

Cube each side. $\qquad\qquad\qquad\qquad\qquad L^3 = L + 1$

$$\Rightarrow \quad L^3 - L - 1 = 0$$

Example 6

(a) Show that the equation $x^3 - x^2 - 5x - 1 = 0$ can be rearranged into
$$x = \sqrt{\frac{5x+1}{x-1}}$$

(b) Use the rearranged form to write a recurrence relation for the purpose of obtaining a solution to the original equation.

(c) With $x_0 = 3$, find to two decimal places the limit to which values from the recurrence relation tend.

Solution

(a)

$$x^3 - x^2 - 5x - 1 = 0$$
$$\Rightarrow \quad x^3 - x^2 = 5x + 1$$
$$\Rightarrow \quad x^2(x - 1) = 5x + 1$$

Dividing both sides by $x - 1$, $\qquad x^2 = \dfrac{5x+1}{x-1}$

Taking the positive square root of both sides, $\quad x = \sqrt{\dfrac{5x+1}{x-1}}$

> *If you cannot see how to start, first try working backwards*
> from $x = \sqrt{\dfrac{5x+1}{x-1}}$
> to $x^3 - x^2 - 5x - 1 = 0$.

(b) $x_{n+1} = \sqrt{\dfrac{5x_n + 1}{x_n - 1}}$

(c) Limit = 2.87 to 2 d.p.

Exercise B (answers p 201)

1 For each of these recurrence relations, find x_1, x_2 and x_3 to four decimal places.

(a) $x_0 = 1.4$, $x_{n+1} = \ln(x_n + 3)$

(b) $x_0 = 1.5$, $x_{n+1} = \sqrt{x_n + 1}$

(c) $x_0 = 0.5$, $x_{n+1} = 2 - \dfrac{x_n^3}{8}$

(d) $x_0 = 0.4$, $x_{n+1} = \dfrac{e^{x_n}}{4}$

2 Match each equation to a possible iterative formula.

(a) $x^2 - 8x + 1 = 0$

(b) $x^2 - 8x + 8 = 0$

(c) $x^2 - 8x - 1 = 0$

P $\quad x_{n+1} = \dfrac{1}{x_n} + 8$

Q $\quad x_{n+1} = \sqrt{8x_n - 1}$

R $\quad x_{n+1} = \dfrac{x_n^2}{8} + 1$

3 Each of the iterative formulae below produces a sequence that converges to a limit. For each one,

    (i) find x_1, x_2 and x_3 to three decimal places and continue the sequence to obtain the limit correct to three decimal places

    (ii) write an equation in the form $f(x) = 0$ for which the formula provides an approximate solution

(a) $x_0 = 0.6$, $\ x_{n+1} = \cos(x_n)$

(b) $x_0 = 1.5$, $\ x_{n+1} = \dfrac{1}{x_n} + 1$

(c) $x_0 = 3$, $\ x_{n+1} = \frac{1}{3}(2^{x_n})$

(d) $x_0 = 0.1$, $\ x_{n+1} = \frac{1}{2}(1 - x_n^2)$

4 (a) A function is defined as $f(x) = x^3 + 2x - 1$.

Show that there is a root α of $f(x) = 0$ in the interval $(0, 1)$.

(b) Show that $f(x) = 0$ can be rearranged as

$$x = \frac{1 - x^3}{2}$$

(c) The root α is to be estimated using the iterative formula

$$x_{n+1} = \frac{1 - x_n^3}{2}$$

Taking $x_0 = 0$, apply the formula to obtain x_1, x_2 and x_3 to four decimal places.

(d) By considering the sign of $f(x)$ for suitable values of x, show that $\alpha = 0.4534$ to four decimal places.

5 Each iterative formula below leads to a limit. Show that in each case the limit satisfies the equation $2x^2 - 5x + 1 = 0$.

(a) $x_{n+1} = \dfrac{1 + 2x_n^2}{5}$

(b) $x_{n+1} = \dfrac{1}{5 - 2x_n}$

(c) $x_{n+1} = \sqrt{\dfrac{5x_n - 1}{2}}$

(d) $x_{n+1} = \frac{1}{2}\left(5 - \dfrac{1}{x_n}\right)$

6 (a) An iterative formula is defined as follows.

$$x_{n+1} = \frac{1}{\sqrt[3]{x_n + 1}}$$

Using $x_0 = 0.8$, find x_1, x_2 and x_3 to six decimal places.

(b) Given that the limit of of the sequence is k, show that k satisfies the equation $k^4 + k^3 - 1 = 0$.

7 An approximate solution to the equation $x^3 - 5x - 3 = 0$ may be obtained using

an iterative formula of the form $x_{n+1} = \sqrt{\dfrac{p}{x_n} + q}$, where p and q are constants.

Find the values of p and q.

8 The graphs of $y = \sin x + \frac{1}{2}$ and $y = x$ are shown sketched.

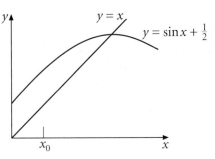

Trace the sketch and draw a cobweb or staircase diagram to show the convergence of the recurrence relation

$$x_0 = 0.5, \quad x_{n+1} = \sin x_n + \tfrac{1}{2}$$

9 (a) What, to the nearest integer, is the solution to $x^3 = 10$?

(b) Show that the equation can be arranged into the form $x = \sqrt{\dfrac{10}{x}}$.

Using the corresponding iterative formula with your answer to part (a) as the starting value, obtain the solution to $x^3 = 10$ to three decimal places.

(c) Show that $x^3 = 10$ can also be arranged into $x = \sqrt{\sqrt{10x}}$.

Using the new corresponding iterative formula with the same starting value, obtain the positive solution to $x^3 = 10$ to three decimal places.

(d) Which formula gives the solution more quickly?

10 (a) By sketching appropriate graphs and substituting suitable integer values of x, find an interval that contains the root of

$$x^2 - 1 = 6\sqrt{x}$$

(b) Show that $x = \sqrt{6\sqrt{x} + 1}$ is a rearrangement of this equation.

(c) Choosing a suitable starting value, solve the equation by an iterative method, giving your answer to three decimal places.

11 (a) Show that $x^3 + x^2 - 3x - 1 = 0$ can be rearranged as

$$x = \sqrt{\dfrac{3x + 1}{x + 1}}$$

(b) The equation $x^3 + x^2 - 3x - 1 = 0$ has only one positive root α.

The iteration formula $x_{n+1} = \sqrt{\dfrac{3x_n + 1}{x_n + 1}}$ may be used to find an approximation to α.

Taking $x_0 = 3$, find to four decimal places the values of x_1, x_2 and x_3.

(c) Giving your reasons, write down a value of x_0 for which the iteration formula above does not produce a valid value for x_1.

*12 (a) For each of these recurrence relations, find x_1, x_2, x_3, x_4.

(i) $x_0 = 2, \quad x_{n+1} = x_n^3 - 1$ **(ii)** $x_0 = 0.4, \quad x_{n+1} = \dfrac{1}{x_n^2}$

(b) What happens?

(c) In each case, carefully draw a staircase or cobweb diagram to show the situation.

(d) What feature of the curved graphs in your diagrams results in a limit not being approached?

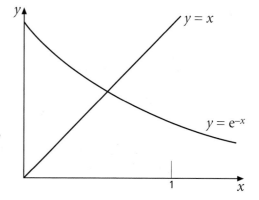
Mixed questions (answers p 203)

1 The graphs of $y = x^3 + 1$ and $y = 2e^{-x}$ intersect at point A.

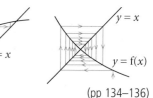

(a) Show that the x-coordinate of A satisfies the equation

$$x^3 - 2e^{-x} + 1 = 0$$

(b) Show that the x-coordinate of A lies between 0.5 and 0.6.

2 The graphs of $y = e^{-x}$ and $y = x$ are shown sketched.

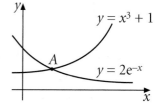

(a) Trace the sketch and draw a cobweb or staircase diagram to show the convergence of the recurrence relation

$$x_0 = 1, \quad x_{n+1} = e^{-x_n}$$

(b) Find the limit to which the sequence tends, to 2 d.p.

3 The solution of the equation $f(x) = 0$, where $f(x) = x + \ln 3x - 5$, may be estimated using the iterative formula $x_{n+1} = 5 - \ln 3x_n$ with $x_0 = 3$.

(a) By repeated application of the formula, obtain an estimate of the solution to an accuracy of four significant figures.

(b) By considering the change of sign of $f(x)$ over a suitable interval, justify the accuracy of your answer to part (a).

4 (a) Without using a calculator, show that the equation $x^3 - 12 = 0$ has a root in the interval $2 \leq x \leq 4$.

(b) (i) Show that the equation $x = \dfrac{3x}{4} + \dfrac{3}{x^2}$ can be arranged to give the equation

$$x^3 - 12 = 0$$

(ii) Use the iterative formula $x_{n+1} = \dfrac{3x_n}{4} + \dfrac{3}{x_n^2}$, starting with $x_0 = 4$,

to find the values of x_1, x_2 and x_3, giving your answers to two decimal places.

(iii) The graphs of $y = \dfrac{3x}{4} + \dfrac{3}{x^2}$ and $y = x$ are shown below.

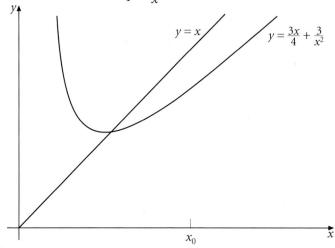

Trace the graphs and draw a staircase diagram to show the convergence of the sequence $x_0, x_1, x_2, \ldots$

(iv) Write down the **exact** value of the limit of this sequence.

5 A function is defined as $f(x) = x^3 - \dfrac{x^2}{2} - 5$.

(a) The equation $f(x) = 0$ has one real root α. Show that α lies in the interval $(1.5, 2)$.

(b) Show that $f(x) = 0$ can be rearranged as

$$x = \sqrt{\dfrac{x}{2} + \dfrac{5}{x}}$$

(c) α may be estimated using the iterative formula

$$x_{n+1} = \sqrt{\dfrac{x_n}{2} + \dfrac{5}{x_n}}$$

Taking $x_0 = 1.5$, calculate the values of x_1, x_2 and x_3, giving your answers to three decimal places.

(d) Demonstrate that, correct to four decimal places, α is 1.8939.

6 Stating the values of the constants a, b and c, use an iteration of the form

$$x_{n+1} = (ax_n + b)^{\frac{1}{c}} \text{ with } x_0 = 2$$

to find the value of a root of the equation $x^4 - 2x - 3 = 0$, correct to 4 d.p.

Test yourself (answers p 203)

1 Show that the equation $\sqrt{x} = 2\ln x$ has a root between 2 and 2.1.

2 (a) On the same axes, sketch the graphs of $y = \ln x$ and $y = x^3 - 2$.

    (b) Using your sketch, write down how many roots the equation $\ln x = x^3 - 2$ has.

    (c) Show that the equation $\ln x = x^3 - 2$ has a root between 1.3 and 1.4.

3 A sequence is defined by $x_{n+1} = \sqrt{x_n + 6}$, $x_0 = 1$.

    (a) Find the values of $x_1, x_2,$ and x_3, rounding to four significant figures.

    (b) Given that the limit of this sequence is K,

        (i) show that K satisfies the equation $K^2 - K - 6 = 0$

        (ii) hence find the exact value of K.

4 (a) Sketch, on the same set of axes, the graphs of
$$y = 2 - e^{-x} \text{ and } y = \sqrt{x}.$$
    (It is not necessary to find the coordinates of any points of intersection with the axes.)

    Given that $f(x) = e^{-x} + \sqrt{x} - 2$, $x \geq 0$,

    (b) explain how your graphs show that the equation $f(x) = 0$ has only one solution,

    (c) show that the solution of $f(x) = 0$ lies between $x = 3$ and $x = 4$.

    The iterative formula $x_{n+1} = (2 - e^{-x_n})^2$ is used to solve the equation $f(x) = 0$.

    (d) Taking $x_0 = 4$, write down the values of x_1, x_2, x_3 and x_4, and hence find an approximation to the solution of $f(x) = 0$, giving your answer to three decimal places.   Edexcel

5 Show that $x = \sqrt{ax^3 - 3}$ may be rearranged into the form $x^3 - 4x^2 - 12 = 0$, where a is a constant to be found.

6 $$f(x) = x^3 + x^2 - 4x - 1$$

    The equation $f(x) = 0$ has only one positive root, α.

    (a) Show that $f(x) = 0$ can be rearranged as
$$x = \sqrt{\frac{4x+1}{x+1}}, \ x \neq -1$$

    The iteration formula $x_{n+1} = \sqrt{\dfrac{4x_n+1}{x_n+1}}$ is used to find an approximation to α.

    (b) Taking $x_1 = 1$, find, to two decimal places, the values of x_2, x_3 and x_4.

    (c) By choosing values of x in a suitable interval, prove that $\alpha = 1.70$, correct to two decimal places.

    (d) Write down a value of x_1 for which the iteration formula $x_{n+1} = \sqrt{\dfrac{4x_n+1}{x_n+1}}$ does not produce a valid value for x_2.

    Justify your answer.   Edexcel

10 Proof

A Introducing proof (answers p 205)

Most mathematicians would agree that proof is at the heart of mathematics. A proof uses known truths and logical argument to show that a statement is true and, once the statement has been proved this way, its truth is certain.

A proof may do more than show the truth of a statement: it may reveal something that gives a better understanding of **why** the statement is true. Many mathematics teachers think that proof is important, not just to establish certainty but to **explain**, helping students to develop a stronger and deeper understanding of ideas and techniques.

A proof must be valid; but whether or not a person finds it convincing depends to a large extent on their mathematical knowledge and understanding: what convinces one person may not convince another.

A1 How would you convince a 10-year-old child that when you add any two odd numbers together you always get an even number?

How would you respond to the question in a mathematics examination 'Prove that the sum of any two odd numbers is always an even number'?

A2 Here are three responses to the question
'Prove that $x^2 - 1 = (x + 1)(x - 1)$ for all values of x.'

Comment on each response.

Response 1

When $x = 1$, $\quad x^2 - 1 = 1^2 - 1 = 0$
$\qquad\qquad\quad (x + 1)(x - 1) = (1 + 1)(1 - 1) = 2 \times 0 = 0$

When $x = 2$, $\quad x^2 - 1 = 2^2 - 1 = 3$
$\qquad\qquad\quad (x + 1)(x - 1) = (2 + 1)(2 - 1) = 3 \times 1 = 3$

When $x = 3$, $\quad x^2 - 1 = 3^2 - 1 = 8$
$\qquad\qquad\quad (x + 1)(x - 1) = (3 + 1)(3 - 1) = 4 \times 2 = 8$

When $x = 4$, $\quad x^2 - 1 = 4^2 - 1 = 15$
$\qquad\qquad\quad (x + 1)(x - 1) = (4 + 1)(4 - 1) = 5 \times 3 = 15$

… and so on.

Response 2

$(x + 1)(x - 1) = (x \times x) + (x \times -1) + (1 \times x) + (1 \times -1)$
$\qquad\qquad\qquad = x^2 - x + x - 1$
$\qquad\qquad\qquad = x^2 - 1$

D   **Response 3**

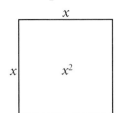
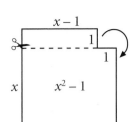
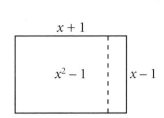

K $x^2 - 1 = (x + 1)(x - 1)$ is true for **all** values of x and so is an example of an **identity**.

Hence we can use the **identity symbol** $\equiv$ to write it as $x^2 - 1 \equiv (x + 1)(x - 1)$.

B Disproof by counterexample

A **conjecture** is a statement that has not yet been proved true or false.

It is sometimes straightforward to prove a conjecture is false.
For example, we can show that the conjecture 'All multiples of 3 are odd' is
false by producing one multiple of 3 that is even, for example 6.
This is called a **counterexample** and only one counterexample is needed
to prove a conjecture is false.

Example 1

By finding a suitable counterexample, prove that the conjecture below is false.
$$a^3 + b^3 \equiv (a + b)^3$$

Solution

When $a = 4$ and $b = 1$, then $a^3 + b^3 = 4^3 + 1^3 = 64 + 1 = 65$

and $(a + b)^3 = (4 + 1)^3 = 5^3 = 125$

Clearly $65 \neq 125$, a counterexample that shows the conjecture is false.

Exercise B (answers p 205)

Prove that each of these conjectures is false by finding a suitable counterexample.

1 All prime numbers are odd.

2 $\dfrac{\cos(3\theta)}{3} \equiv \cos\theta$

3 The product of two odd numbers is always a multiple of 3.

4 $\cos(-\theta) \equiv -\cos\theta$

5 $\sin(2\theta) \equiv 2\sin\theta$

6 For all real values of x, $x^2 \geq x$.

7 $(a + b)^2 \equiv a^2 + b^2$

8 The function $f(x) = \operatorname{cosec} x$ is defined for all real values of x.

9 The product of two different irrational numbers is always irrational.

10 $\sin A + \sin B \equiv \sin (A + B)$

11 $\sqrt{x + y} \equiv \sqrt{x} + \sqrt{y}$

12 $\dfrac{1}{2 \cos x} = 2 \sec x$ for all values of x for which $\cos x \neq 0$.

13 $e^{2x} \equiv e^x + e^2$

14 For all positive real values of A and B, $\ln (AB) = \ln A \ln B$.

15 For all real values of x, $|x + 2| < 3 \Rightarrow |x| < 1$.

16 For all real values of x and y, $x^2 > y^2 \Rightarrow x > y$.

C Constructing a proof (answers p 206)

Students often find producing a mathematical proof extremely difficult. They find it hard to know where to start or else get stuck in the middle. Sometimes they can follow a proof that someone else has produced but feel that they could never have constructed it for themselves.

The truth is that many published proofs are the polished result of a messy process that has involved many dead-ends. The final proof may look effortless and elegant but does not reveal the work involved in constructing it. Most people only feel confident about proof after a lot of experience of proving things in a variety of mathematical contexts.

Consider this conjecture.   For all odd numbers n, $n^2 - 1$ is divisible by 8.

The following shows how one person proved the conjecture true.

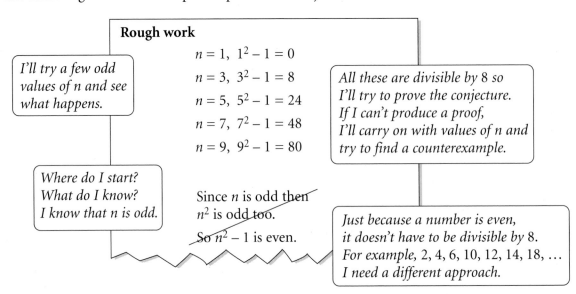

> *All odd numbers can be written as $2 \times$ some number + 1.*
> *For example, $11 = 2 \times 5 + 1$.*
>
> *So I can write n as $2k + 1$.*

$$n = 2k + 1$$

$$\text{so } n^2 = (2k + 1)^2$$

$$= 4k^2 + 4k + 1$$

$$\text{so } n^2 - 1 = 4k^2 + 4k$$

$$= 4k(k + 1)$$

> *Clearly this is divisible by 4 but I need to show it is divisible by 8. I'll factorise and see if that helps.*

> *If I could show that $k(k + 1)$ was even then the number would be the product of 4 and an even number, which is divisible by 8. But I see that k and $k + 1$ are consecutive integers so one of them must be even and hence the product must be even. I think I'm there!*

Final proof

As n is odd there is some integer k such that $n = 2k + 1$.

So $n^2 - 1 = (2k + 1)^2 - 1$

$$= 4k^2 + 4k + 1 - 1$$

$$= 4k^2 + 4k$$

$$= 4k(k + 1)$$

Since k and $k + 1$ are consecutive integers, one of them must be even, so the product $k(k + 1)$ is even too.

So $4k(k + 1)$ is the product of 4 and an even number.
Hence $4k(k + 1)$ is divisible by 8.

So for all odd n, $n^2 - 1$ is divisible by 8.

C1 What mathematical experience, skills and knowledge do you think a reader needs to be convinced by the above proof?

C2 One of these conjectures is true and one of them is false.
- For all odd numbers n, $n^2 + 3$ is divisible by 4.
- For all odd numbers n, $n^2 + 5$ is divisible by 3.

Decide which conjecture is true and prove it.

D Direct proof

The proof outlined in section C is an example of a direct proof: it starts with what is known and proceeds, by a sequence of logical steps, to the conclusion.

Most of the proofs that you construct as part of your A2 studies are of this type.

Example 2

Prove $\cos\theta\cot\theta + \sin\theta \equiv \operatorname{cosec}\theta$ and find any values for which the identity is not defined.

Solution

Use the identity $\cot\theta \equiv \dfrac{\cos\theta}{\sin\theta}$.

$$\cos\theta\cot\theta + \sin\theta \equiv \cos\theta \times \frac{\cos\theta}{\sin\theta} + \sin\theta$$

$$\equiv \frac{\cos^2\theta}{\sin\theta} + \sin\theta$$

Write $\sin\theta$ as $\dfrac{\sin^2\theta}{\sin\theta}$.

$$\equiv \frac{\cos^2\theta}{\sin\theta} + \frac{\sin^2\theta}{\sin\theta}$$

Add.

$$\equiv \frac{\cos^2\theta + \sin^2\theta}{\sin\theta}$$

Use the identity $\cos^2\theta + \sin^2\theta \equiv 1$.

$$\equiv \frac{1}{\sin\theta}$$

$$\equiv \operatorname{cosec}\theta$$

The identity is not defined for any values that give 0 on a denominator.

$\cot\theta \equiv \dfrac{\cos\theta}{\sin\theta}$ and $\operatorname{cosec}\theta \equiv \dfrac{1}{\sin\theta}$.

The denominator $\sin\theta$ is 0 for $\theta = 0, \pm\pi, \pm2\pi, \pm3\pi, \ldots$

So the identity is not defined for $\theta = 0, \pm\pi, \pm2\pi, \pm3\pi, \ldots$

We can write this as $\theta \neq n\pi,\ n \in \mathbb{Z}$. $\mathbb{Z}$ is shorthand for the set of integers.

Example 3

Show that the cube of any even number is divisible by 8.

Solution

Any even number can be written as $2k$ for some integer k.

The cube of this number is $(2k)^3$ which is equivalent to $8k^3$.

Since k is an integer then k^3 is an integer too and so $8k^3$ must be divisible by 8.

Hence the cube of any even number is divisible by 8.

Exercise D (answers p 206)

Prove each of these.

1 The sum of any two consecutive odd numbers is a multiple of 4.

2 The area of a trapezium is $\frac{1}{2}h(a + b)$, where h is the distance between the parallel edges and a and b are the lengths of the parallel edges.

3 $\cos^2\theta + \sin^2\theta \equiv 1$

4 The sum of the first n odd numbers is n^2.

5 For any four consecutive integers, the difference between the product of the last two and the product of the first two of these numbers is equal to their sum.

6 The graphs of $y = |\sin x \cos x + 1|$ and $y = \sin x \cos x + 1$ are identical.

7 $\sin\theta \tan\theta \equiv \sec\theta - \cos\theta$, $\theta \neq \dfrac{n\pi}{2}$, $n \in \mathbb{Z}$

8 (a) If $f(x) = \left(1 - \frac{1}{2}x^2\right)\cos x + x\sin x$ then $f'(x) = \frac{1}{2}x^2\sin x$.

   (b) Hence write down the indefinite integral $\int x^2 \sin x \, dx$.

9 The product of four consecutive integers is always 1 less than a perfect square.

*10 $\dfrac{\tan^2 A + \cos^2 A}{\sin A + \sec A} \equiv \sec A - \sin A$, $\theta \neq \dfrac{n\pi}{2}$, $n \in \mathbb{Z}$

*11 $k^3 - k$ is divisible by 6 for all $k \in \mathbb{Z}$.

*12 $9^n - 1$ is a multiple of 8 for any positive integer n.

*13 If p is a prime number such that $p > 3$, then $p^2 - 1$ is a multiple of 24.

E Proof by contradiction

For this method of proof, assume that what you want to prove is in fact false and try to derive a logical contradiction.

One of the best known proofs of this kind is shown below.

Proof by contradiction that $\sqrt{2}$ is irrational

Assume that $\sqrt{2}$ is rational.

Then there must exist integers p and q such that $\sqrt{2} = \dfrac{p}{q}$ in its simplest form.

$$\sqrt{2} = \frac{p}{q} \implies 2 = \frac{p^2}{q^2}$$
$$\implies p^2 = 2q^2$$

So p^2 is even and therefore p is even too.

If p is even then there exists an integer k such that $p = 2k$ and so $p^2 = (2k)^2 = 4k^2$.

Hence $2q^2 = 4k^2$ which gives $q^2 = 2k^2$ and so q^2 is even and therefore q is even too.

But if both p and q are even, then $\dfrac{p}{q}$ is not in its simplest form and our original assumption was false.

We conclude that $\sqrt{2}$ is irrational.

One of the first known proofs by contradiction is Euclid's proof that there are infinitely many prime numbers. It is considered to be one of the most elegant proofs in mathematics.

Proof by contradiction that there are infinitely many primes

Assume that there are a finite number of primes.

Let $p_1 = 2, p_2 = 3, p_3 = 5, p_4 = 7, \ldots$ be the primes in ascending order and let p_n be the largest.

Now consider the integer N which is one more than the product of all the primes:

$$N = p_1 p_2 p_3 p_4 \cdots p_n + 1$$

This number is not divisible by any of the primes $p_1, p_2, p_3, \ldots p_n$ as division by each leaves a remainder of 1. Hence N is divisible by a prime larger than p_n or is itself prime.

This contradicts the assumption that p_n is the largest prime.

Hence our original assumption is false and there must be infinitely many primes.

Example 4

Prove that the equation $x^3 = 99x + 1$ has no integer solutions.

Solution

Assume that the equation $x^3 = 99x + 1$ does have at least one integer solution and call it k. Hence $k^3 = 99k + 1$.

Now k is either even or odd.

If k is even then k^3 is even and $99k + 1$ is odd which is not possible.
If k is odd then k^3 is odd and $99k + 1$ is even which is not possible either.

In both cases we have a contradiction.
Hence our original assumption is false and the equation $x^3 = 99x + 1$ has no integer solutions.

Exercise E (answers p 208)

1 Prove each of these by contradiction.

(a) $\sqrt[3]{2}$ is irrational

(b) $x^4 = 45x + 1$ has no integer solutions.

(c) There are no pairs of positive integers x and y such that $x^2 - y^2 = 10$.

2 A teacher calls Jo and Raj to the front of the class and shows them two £5 notes and a £50 note.

She asks them to shut their eyes and raise a hand over their head.
The teacher places a £5 note in each of their hands.
She hides the £50 note and tells them to open their eyes.

Each can then see the other's note but not their own.

The teacher explains that the first person to say which note is in their own hand and prove it will win the £50 note.

After a few minutes silence, Jo proves that she must be holding a £5 note.
How do you think she did it?

F Convincing but flawed

A proof must be convincing, but not all arguments that look convincing are valid proofs.

Exercise F (answers p 208)

Explain what is wrong with each of the following arguments.

1 Statement $2 = 1$

    Argument   Suppose that $a = b$

                 Hence $a^2 = ab$

$$\Rightarrow \qquad a^2 + a^2 = ab + a^2$$
$$\Rightarrow \qquad 2a^2 = ab + a^2$$
$$\Rightarrow \qquad 2a^2 - 2ab = ab + a^2 - 2ab$$
$$\Rightarrow \qquad 2a^2 - 2ab = a^2 - ab$$
$$\Rightarrow \qquad 2(a^2 - ab) = a^2 - ab$$

                 and dividing both sides by $(a^2 - ab)$ gives $2 = 1$ as required.

2 Statement   The sum of any two numbers is even.

    Argument   Let n represent any number.

                 Then the sum of any two numbers is $n + n$ which is $2n$.

                 $2n$ is even so the sum of any two numbers is even.

3 Statement $0 = 2$

    Argument   We know that $\cos^2\theta + \sin^2\theta \equiv 1$

                 Hence $\cos^2\theta \equiv 1 - \sin^2\theta$

$$\Rightarrow \qquad \cos\theta \equiv \sqrt{1 - \sin^2\theta}$$
$$\Rightarrow \qquad 1 + \cos\theta \equiv 1 + \sqrt{1 - \sin^2\theta}$$

                 Now this is true for all values of θ, so substitute $\theta = \pi$ to obtain

$$1 + \cos\pi = 1 + \sqrt{1 - \sin^2\pi}$$
$$\Rightarrow \qquad 1 + {-1} = 1 + \sqrt{1 - 0^2}$$
$$\Rightarrow \qquad 0 = 2$$

4 Statement $n^2 - n + 41$ is prime for all $n \in \mathbb{Z}$.

    Argument   I have tested the statement for all integers from -30 to 30 and $n^2 - n + 41$ always produced a prime number.

5 Statement The diagram below is produced by joining each vertex of a square to the mid-point of each edge.

The shape *ABCDEFGH* is a regular octagon.

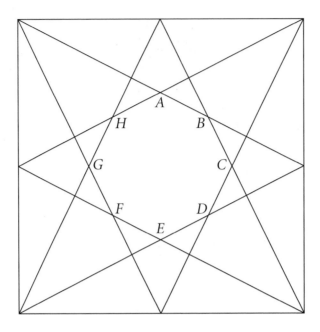

Argument The whole figure is symmetrical about the lines *AE*, *BF*, *CG* and *DH* so all the edges of *ABCDEFGH* are equal. Hence the shape is a regular octagon.

6 Statement The maximum number of regions that can be formed in a circle by drawing all possible chords between n points on the circumference is 2^{n-1}.

Argument

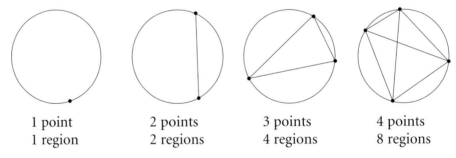

1 point   2 points   3 points   4 points
1 region   2 regions   4 regions   8 regions

Clearly, the pattern continues, doubling each time. So we are looking for the nth term of the geometric sequence 1, 2, 4, 8, …, which is 2^{n-1}.

7 Statement   This black shape ■ has an area of 0 square units.

Argument   Two right-angled triangles, three rectangles and a square are arranged to make a right-angled triangle with an area of $\frac{1}{2}(13 \times 8) = 52$ square units.

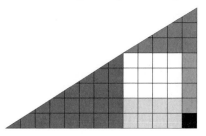

The same pieces are arranged to make another triangle with the same area.

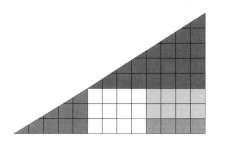

However, the black shape is left over.
So the black shape has an area of 0 square units.

Test yourself (answers p 209)

1 Find a counterexample to show that the following conjecture is false:
'All even numbers greater than 2 can be expressed as the sum of two primes in only one way.'

2 $f(n) = n^2 + n + 1$, where n is a positive integer.

Classify the following statements about $f(n)$ as true or false.
If a statement is true, prove it; if it is false, provide a counterexample.

(a) $f(n)$ is always a prime number   (b) $f(n)$ is always an odd number   Edexcel

3 (a) Prove, by counterexample, that the statement
  '$\sec(A + B) \equiv \sec A + \sec B$, for all A and B' is false.

(b) Prove that $\tan\theta + \cot\theta \equiv 2\csc 2\theta$, $\theta \neq \dfrac{n\pi}{2}$, $n \in \mathbb{Z}$.   Edexcel

4 Prove each of these.

(a) For any four consecutive integers, the difference between the product of the first and last and the product of the middle two of these integers is always 2.

(b) The sum of any three consecutive multiples of 4 is a multiple of 12.

(c) When a is positive, $\ln a^3 - \ln a^2 - \ln a \equiv 0$.

Answers

1 Rational expressions and division

A Simplifying (p 6)

A1 (a) (i) $\frac{1}{2}$ (ii) $\frac{1}{5}$ (iii) $\frac{1}{11}$

(b) (i) $\frac{1}{101}$, following the pattern in part (a)

(ii) $f(100) = \dfrac{100+3}{100^2 + 4 \times 100 + 3} = \dfrac{103}{10\,403} = \dfrac{1}{101}$

(c) (i) $\dfrac{1}{k+1}$, following the pattern

(ii) $f(k) = \dfrac{k+3}{k^2 + 4k + 3} = \dfrac{k+3}{(k+1)(k+3)} = \dfrac{1}{k+1}$

(d) $f(n) = \dfrac{1}{n+1}$. For all positive integers $n + 1 \geq 2$ so $f(n) \leq \frac{1}{2}$.

A2 (a) (i) $\frac{1}{2}$ (ii) $\frac{4}{5}$ (iii) $\frac{21}{22}$

(b) (i) $\dfrac{k+1}{k+2}$, following the pattern

(ii) $f(k) = \dfrac{k^2 + 6k + 5}{k^2 + 7k + 10} = \dfrac{(k+1)(k+5)}{(k+2)(k+5)} = \dfrac{k+1}{k+2}$

(c) The fraction will always be in its lowest terms, the numerator being one less than the denominator. No integer n satisfies both $n + 1 = 8$ and $n + 2 = 11$. Hence there is no integer solution to $f(n) = \frac{8}{11}$.

A3 The fraction $\dfrac{5x+10}{x^2 + 2x}$ is equivalent to $\dfrac{5(x+2)}{x(x+2)}$

which is equivalent to $\dfrac{5}{x}$.

When x is a multiple of 5 then the fraction will simplify to $\dfrac{1}{k}$ where $x = 5k$.

Exercise A (p 7)

1 (a) $\dfrac{x}{2}$ (b) $\dfrac{x-3}{x+1}$ (c) 3

(d) $\frac{1}{5}$ (e) $\dfrac{3n+2}{2n+1}$

2 (a) $\dfrac{14x - 4x^2}{x-7} = \dfrac{-2x(x-7)}{x-7} = -2x$

(b) (i) -5 (ii) $-n$

3 (a) $x + 3$ (b) $\dfrac{x-1}{2}$ (c) $\dfrac{1}{x+2}$

(d) $\dfrac{x}{x-5}$ (e) $\dfrac{n}{3n+1}$

4 (a) $g(3) = \frac{1}{2}, g(4) = \frac{2}{5}$

(b) $g(n) = \dfrac{2n+18}{n^2 + 10n + 9} = \dfrac{2(n+9)}{(n+1)(n+9)} = \dfrac{2}{n+1}$

When n is odd, $n + 1$ is divisible by 2 and so the fraction will simplify to the unit fraction $\dfrac{1}{k}$ where k is the integer $(n + 1) \div 2$.

(c) $n = 8$

5 (a) $\dfrac{n+4}{n+6}$ (b) $\dfrac{n+7}{n-5}$ (c) $\dfrac{2(n-6)}{n+5}$

(d) $\dfrac{3n-1}{n+2}$ (e) $\dfrac{n-1}{n+2}$ (f) $\dfrac{2n-1}{3(n-4)}$

(g) $\dfrac{1}{n-1}$ (h) $\dfrac{n+8}{n(n+3)}$

6 (a) 6   (b) $2x$ (c) $\dfrac{5(x+4)}{x+2}$

(d) $\dfrac{2(x+5)}{x-2}$ (e) $\dfrac{x+7}{x-1}$ (f) $\dfrac{x-5}{x-1}$

7 (a) $\dfrac{5}{2x}$ (b) $\dfrac{2(x+1)}{x}$ or $2 + \dfrac{2}{x}$

(c) $x + 3$ (d) $\dfrac{3x+2}{2(x-7)}$

8 $\dfrac{1}{x} \div y = \dfrac{1}{x} \div \dfrac{y}{1} = \dfrac{1}{x} \times \dfrac{1}{y} = \dfrac{1}{xy}$

9 (a) $\dfrac{y}{x}$ (b) $\dfrac{1}{3x}$ (c) $2xy$

(d) $\dfrac{8}{x}$ (e) $\dfrac{1}{50x}$

10 $\dfrac{6}{\left(\dfrac{2}{x-1}\right)} = 6 \div \dfrac{2}{x-1} = 6 \times \dfrac{x-1}{2} = 3(x-1)$

11 (a) (i) $f(2) = \frac{1}{2}, g(2) = \frac{1}{8}$ (ii) 4

(b) $\dfrac{f(x)}{g(x)} = \dfrac{6x+8}{3x^2 + 10x + 8} \div \dfrac{2}{x^2 + 4x + 4}$

$= \dfrac{6x+8}{3x^2 + 10x + 8} \times \dfrac{x^2 + 4x + 4}{2}$

$= \dfrac{2(3x+4)}{(3x+4)(x+2)} \times \dfrac{(x+2)(x+2)}{2}$

$= x + 2$

Since x is an integer, $x + 2$ is too.

B Adding and subtracting

Exercise B (p 10)

1 (a) $\dfrac{3(x+1)}{x(2x+3)}$ **(b)** $\dfrac{5}{x(x+5)}$ **(c)** $\dfrac{13x-4}{3x(x-1)}$

(d) $\dfrac{2}{x(x-1)}$

2 (a) $\dfrac{x^2+5}{x}$ **(b)** $\dfrac{6(x+2)}{2x+1}$ **(c)** $\dfrac{4(5-x)}{x-4}$

(d) $\dfrac{3x(1-x)}{3x-2}$

3 (a) $\dfrac{a+b}{ab}$ **(b)** $\dfrac{2y-x}{xy}$ **(c)** $\dfrac{ab+3}{b}$

(d) $\dfrac{a-bc}{c}$ **(e)** $\dfrac{3b-2a}{6ab}$

4 $\dfrac{a}{b(b+1)}$

5 (a) $\dfrac{2x}{(x+1)(x-1)}$ or $\dfrac{2x}{x^2-1}$

(b) $\dfrac{5x-1}{(3x+1)(x+3)}$ **(c)** $\dfrac{x^2-7}{(x-2)(x-5)}$

(d) $\dfrac{20}{(x+4)(x-1)}$ **(e)** $\dfrac{3x^2-13}{(x+3)(2x-1)}$

(f) $\dfrac{48}{(x+1)(x-7)}$

6 (a) $\dfrac{3(2x-7)}{x(x-4)}$ **(b)** $\dfrac{x-3}{(x-1)(x+3)}$

(c) $\dfrac{2x+13}{(x+4)(x+7)}$ **(d)** $\dfrac{8}{(x-7)(x+1)(2x-1)}$

(e) $\dfrac{c+a}{abc}$ **(f)** $\dfrac{4z^2-3x^2}{12xyz}$

7 (a) $\dfrac{4}{(2x+1)(x-1)}+\dfrac{12}{(2x+1)(x+3)}$

(b) $\dfrac{4}{(2x+1)(x-1)}+\dfrac{12}{(2x+1)(x+3)}$

$=\dfrac{4(x+3)+12(x-1)}{(2x+1)(x-1)(x+3)}$

$=\dfrac{4x+12+12x-12}{(2x+1)(x-1)(x+3)}$

$=\dfrac{16x}{(2x+1)(x-1)(x+3)}$

8 (a) $\dfrac{2(6x+1)}{(x+5)(2x-1)}$ **(b)** $\dfrac{4(x+3)}{x(x+2)(x+4)}$

(c) $\dfrac{x+4}{x(x-3)}$ **(d)** $\dfrac{x-3}{(x+1)(x-1)}$ or $\dfrac{x-3}{x^2-1}$

9 $\dfrac{x-6}{x-5}$

10 (a) $\frac{6}{5}$

(b) First show that $f(x)=\dfrac{x+1}{x}=1+\dfrac{1}{x}$

When $x>0$ then $\dfrac{1}{x}>0$ so $1+\dfrac{1}{x}>1$ giving

$f(x)>1$ as required.

11 (a) $\dfrac{x+2}{2x}$ **(b)** $\dfrac{2x}{x+2}$

12 $\dfrac{1}{\dfrac{3}{x-4}-1}=\dfrac{1}{\dfrac{3-(x-4)}{x-4}}=\dfrac{1}{\dfrac{7-x}{x-4}}=\dfrac{x-4}{7-x}$

C Extension: Leibniz's harmonic triangle (p 12)

C1 The nth term is the reciprocal of the nth triangle number $\frac{1}{2}n(n+1)$ which is $\dfrac{1}{\frac{1}{2}n(n+1)}=\dfrac{2}{n(n+1)}$ (multiplying top and bottom by 2).

C2 (a) $\dfrac{2}{n}-\dfrac{2}{n+1}=\dfrac{2(n+1)-2n}{n(n+1)}=\dfrac{2n+2-2n}{n(n+1)}$

$=\dfrac{2}{n(n+1)}$ as required

(b) C1 and C2 (a) show that the nth term in the series can be written as the difference of the two fractions $\dfrac{2}{n}$ and $\dfrac{2}{n+1}$. So the first term is $\frac{2}{1}-\frac{2}{2}$, the second term is $\frac{2}{2}-\frac{2}{3}$, the third term is $\frac{2}{3}-\frac{2}{4}$ and so on. Hence the whole series can be written as

$\left(\frac{2}{1}-\frac{2}{2}\right)+\left(\frac{2}{2}-\frac{2}{3}\right)+\left(\frac{2}{3}-\frac{2}{4}\right)+\left(\frac{2}{4}-\frac{2}{5}\right)+\ldots+\left(\dfrac{2}{n}-\dfrac{2}{n+1}\right)$.

(c) The series can be written with the terms grouped as

$$\tfrac{2}{1} + \left(-\tfrac{2}{2} + \tfrac{2}{2}\right) + \left(-\tfrac{2}{3} + \tfrac{2}{3}\right) + \left(-\tfrac{2}{4} + \tfrac{2}{4}\right) + \left(-\tfrac{2}{5} + \tfrac{2}{5}\right)$$

$$+ \ldots + \left(-\frac{2}{n} + \frac{2}{n}\right) - \frac{2}{n+1} \text{ so that the sum of}$$

each bracketed pair is 0. Hence the sum of the whole series is $2 - \dfrac{2}{n+1}$. As a single

fraction this is $\dfrac{2(n+1)-2}{n+1} = \dfrac{2n+2-2}{n+1} = \dfrac{2n}{n+1}$.

(d) The formula gives $\tfrac{10}{6} = \tfrac{5}{3}$.

Adding gives $\tfrac{1}{1} + \tfrac{1}{3} + \tfrac{1}{6} + \tfrac{1}{10} + \tfrac{1}{15}$

$= 1 + \dfrac{10+5+3+2}{30} = 1 + \tfrac{20}{30} = 1\tfrac{2}{3} = \tfrac{5}{3}$ so the

formula gives the correct result.

C3 In C2 (c) it was shown that the sum of the series is $2 - \dfrac{2}{n+1}$.

As n gets larger and larger, $\dfrac{2}{n+1}$ gets closer and

closer to 0 so the sum $2 - \dfrac{2}{n+1}$ gets closer and closer to 2.

Exercise C (p 12)

1 $\tfrac{1}{20} + \tfrac{1}{30} = \dfrac{3+2}{60} = \tfrac{5}{60} = \tfrac{1}{12}$

2 $\tfrac{1}{6}, \tfrac{1}{30}, \tfrac{1}{60}, \tfrac{1}{60}, \tfrac{1}{30}, \tfrac{1}{6}$

3 $\tfrac{1}{90}$

4 (a) The first fraction in diagonal 1 can be written as $\tfrac{1}{1}$. We know that the denominators in the first diagonal increase by 1 so the first fraction and its successor are $\tfrac{1}{1}$ and $\tfrac{1}{2}$, the second and its successor are $\tfrac{1}{2}$ and $\tfrac{1}{3}$ and so on giving the kth fraction and its successor as $\dfrac{1}{k}$ and $\dfrac{1}{k+1}$.

(b) Each fraction in diagonal 1 is the sum of the two fractions below it. Hence each fraction in diagonal 2 is the difference of the fraction to the left in the row above and the fraction to the left in the same row. So the nth fraction in diagonal 2 is the nth fraction in diagonal 1 minus the $(n + 1)$th fraction in diagonal 1, i.e. $\dfrac{1}{n} - \dfrac{1}{n+1}$.

So in the sum, the first term is $\tfrac{1}{1} - \tfrac{1}{2}$, the second term is $\tfrac{1}{2} - \tfrac{1}{3}$, the third term is $\tfrac{1}{3} - \tfrac{1}{4}$ and so on. Hence the whole series can be written as

$$\left(\tfrac{1}{1} - \tfrac{1}{2}\right) + \left(\tfrac{1}{2} - \tfrac{1}{3}\right) + \left(\tfrac{1}{3} - \tfrac{1}{4}\right) + \left(\tfrac{1}{4} - \tfrac{1}{5}\right) + \ldots$$

$$+ \left(\frac{1}{n} - \frac{1}{n+1}\right).$$

(c) The series can be written with the terms grouped as

$$\tfrac{1}{1} + \left(-\tfrac{1}{2} + \tfrac{1}{2}\right) + \left(-\tfrac{1}{3} + \tfrac{1}{3}\right) + \left(-\tfrac{1}{4} + \tfrac{1}{4}\right) + \left(-\tfrac{1}{5} + \tfrac{1}{5}\right)$$

$$+ \ldots + \left(-\frac{1}{n} + \frac{1}{n}\right) - \frac{1}{n+1} \text{ so that the sum}$$

of each bracketed pair is 0. Hence the sum of the whole series is $1 - \dfrac{1}{n+1}$. As a single

fraction this is $\dfrac{(n+1)-1}{n+1} = \dfrac{n}{n+1}$.

(d) The formula gives $\dfrac{4}{4+1} = \tfrac{4}{5}$.

Adding gives $\tfrac{1}{2} + \tfrac{1}{6} + \tfrac{1}{12} + \tfrac{1}{20} = \dfrac{30+10+5+3}{60}$

$= \tfrac{48}{60} = \tfrac{4}{5}$ so the formula gives the correct result.

(e) As n gets larger and larger, the sum converges to 1. From part (c) above we see that the sum can be written as $1 - \dfrac{1}{n+1}$. As n gets larger and

larger $\dfrac{1}{n+1}$ gets closer and closer to 0, so the

sum $1 - \dfrac{1}{n+1}$ gets closer and closer to 1.

(f) From part (b) we know that the kth fraction in diagonal 2 is $\dfrac{1}{k} - \dfrac{1}{k+1}$; this is equivalent to

$$\dfrac{(k+1)-k}{k(k+1)} = \dfrac{1}{k(k+1)} \text{ as required.}$$

(g) (i) $\tfrac{1}{420} = \dfrac{1}{20 \times 21}$ and so it is the 20th fraction in diagonal 2.

(ii) Row 21

5 (a) From the result of question 4 (f) we know that the kth fraction in diagonal 2 is $\dfrac{1}{k(k+1)}$ so its successor in diagonal 2 is the $(k+1)$th, which is $\dfrac{1}{(k+1)((k+1)+1)} = \dfrac{1}{(k+1)(k+2)}$.

(b) Each fraction in diagonal 3 is the difference of the fraction to the left in the row above and the fraction to the left in the same row. So the nth fraction in diagonal 3 is the nth fraction in diagonal 2 minus the $(n+1)$th fraction in diagonal 2.

From part (a) this is $\dfrac{1}{n(n+1)} - \dfrac{1}{(n+1)(n+2)}$.

So the sum of the first n fractions can be written as

$\left(\frac{1}{2} - \frac{1}{6}\right) + \left(\frac{1}{6} - \frac{1}{12}\right) + \left(\frac{1}{12} - \frac{1}{20}\right) + \left(\frac{1}{20} - \frac{1}{30}\right) + \dots$
$+ \left(\dfrac{1}{n(n+1)} - \dfrac{1}{(n+1)(n+2)}\right)$.

The series can be written with the terms grouped as

$\frac{1}{2} + \left(-\frac{1}{6} + \frac{1}{6}\right) + \left(-\frac{1}{12} + \frac{1}{12}\right) + \left(-\frac{1}{20} + \frac{1}{20}\right) + \dots$
$+ \left(-\dfrac{1}{n(n+1)} + \dfrac{1}{n(n+1)}\right) - \dfrac{1}{(n+1)(n+2)}$

so that the sum of each bracketed pair is 0. Hence the sum of the whole series is

$\frac{1}{2} - \dfrac{1}{(n+1)(n+2)}$.

(c) As n gets larger and larger, the sum converges to $\frac{1}{2}$. The sum of the first n fractions is $\frac{1}{2} - \dfrac{1}{(n+1)(n+2)}$. As n gets larger and larger $\dfrac{1}{(n+1)(n+2)}$ gets closer and closer to 0 so the sum $\frac{1}{2} - \dfrac{1}{(n+1)(n+2)}$ gets closer and closer to $\frac{1}{2}$.

(d) (i) From part (b) we know that the kth fraction in diagonal 3 is $\dfrac{1}{k(k+1)} - \dfrac{1}{(k+1)(k+2)}$ and this is equivalent to $\dfrac{(k+2)-k}{k(k+1)(k+2)}$

$= \dfrac{2}{k(k+1)(k+2)}$ as required.

(ii) $\frac{1}{660}$

6 Investigations (and proofs) leading to:

- the nth fraction in diagonal m is
$\dfrac{(m-1)!}{n(n+1)(n+2)\dots(n+m-1)}$

- the sum of the first n fractions in diagonal m is
$\dfrac{1}{m-1} - \dfrac{(m-2)!}{(n+1)(n+2)\dots(n+m-1)}$

- as n gets larger and larger the sum of the first n fractions in diagonal m converges to a limit of $\dfrac{1}{m-1}$.

7 One method of proof is as follows. From C1 we know that the reciprocal of the nth triangle number T_n can be written as $\dfrac{2}{n(n+1)}$.

Hence the reciprocal of the $(n+1)$th triangle number T_{n+1} can be written as $\dfrac{2}{(n+1)(n+2)}$.

The sum of these reciprocals is

$\dfrac{2}{n(n+1)} + \dfrac{2}{(n+1)(n+2)}$

which is equivalent to

$\dfrac{2(n+2)+2n}{n(n+1)(n+2)}$

$= \dfrac{4n+4}{n(n+1)(n+2)}$

$= \dfrac{4(n+1)}{n(n+1)(n+2)} = \dfrac{4}{n(n+2)}$

We also have

$\dfrac{4}{T_n + T_{n+1} - 1} = \dfrac{4}{\frac{1}{2}n(n+1) + \frac{1}{2}(n+1)(n+2) - 1}$

$= \dfrac{8}{n(n+1) + (n+1)(n+2) - 2}$

$= \dfrac{8}{n^2 + n + n^2 + 3n + 2 - 2}$

$= \dfrac{8}{2n^2 + 4n} = \dfrac{4}{n(n+2)}$

which is the same as the sum above, as required.

D Extension: the harmonic mean (p 14)

D1 (a) 10 **(b)** 10 **(c)** 10

D2 (a) 9 **(b)** 12 **(c)** 7

D3 (a) $\dfrac{b-a}{a+b}$

(b) Since k is a fraction such that $a + ka = b - kb$ then the harmonic mean of a and b is $a + ka$ (or $b - kb$).

$k = \dfrac{b-a}{a+b}$ so the harmonic mean is

$$a + a\left(\dfrac{b-a}{a+b}\right) = \dfrac{a(a+b) + a(b-a)}{a+b}$$

$$= \dfrac{a^2 + ab + ab - a^2}{a+b} = \dfrac{2ab}{a+b} \text{ as required.}$$

(c) $\dfrac{2 \times 10 \times 90}{10 + 90} = \dfrac{1800}{100} = 18$

D4 (a) Dividing the numerator and denominator by ab gives

$$\dfrac{2ab}{a+b} = \dfrac{2}{\frac{a+b}{ab}} = \dfrac{2}{\frac{a}{ab} + \frac{b}{ab}} = \dfrac{2}{\frac{1}{b} + \frac{1}{a}} = \dfrac{2}{\frac{1}{a} + \frac{1}{b}}$$

as required.

(b) The mean of the reciprocals of a and b is $\dfrac{\frac{1}{a} + \frac{1}{b}}{2}$ so the reciprocal of the mean of the reciprocals is $\dfrac{2}{\frac{1}{a} + \frac{1}{b}}$ which is the harmonic mean (from part (a)).

Exercise D (p 15)

1 (a) Let d miles be the distance from Harton to Monyborough. Then the total distance for the return journey is $2d$ miles.

The time for the outward journey is $\dfrac{d}{x}$ and the time for the return journey is $\dfrac{d}{y}$ so the total time is $\dfrac{d}{x} + \dfrac{d}{y} = d\left(\dfrac{1}{x} + \dfrac{1}{y}\right)$. Hence the average speed is $\dfrac{2d}{d\left(\frac{1}{x} + \frac{1}{y}\right)} = \dfrac{2}{\frac{1}{x} + \frac{1}{y}}$ which is the harmonic mean of x and y as required.

(b) $\dfrac{2}{\frac{1}{30} + \frac{1}{60}} = \dfrac{2}{\frac{1}{20}}$ so the average speed is 40 m.p.h.

2 (a) $\dfrac{2 \times 2 \times 6}{2 + 6} = \dfrac{24}{8} = 3$ so 3 is the harmonic mean of 2 and 6.

(b) The reciprocals in order are $\frac{1}{6}, \frac{1}{3}, \frac{1}{2}$. Now $\frac{1}{3} - \frac{1}{6} = \frac{1}{6}$ and $\frac{1}{2} - \frac{1}{3} = \frac{1}{6}$ so there is a common difference and hence an arithmetic sequence.

(c) q is the harmonic mean of p and r so $q = \dfrac{2pr}{p+r}$. The reciprocals of r, q and p are $\dfrac{1}{r}, \dfrac{p+r}{2pr}$ and $\dfrac{1}{p}$.

Now $\dfrac{p+r}{2pr} - \dfrac{1}{r} = \dfrac{p+r-2p}{2pr} = \dfrac{r-p}{2pr}$ and

$\dfrac{1}{p} - \dfrac{p+r}{2pr} = \dfrac{2r-(p+r)}{2pr} = \dfrac{r-p}{2pr}$ so we have a common difference and hence an arithmetic sequence.

3 One proof is as follows. The diagram can be labelled as follows with h as the height and b as the length of the base of $\triangle PST$.

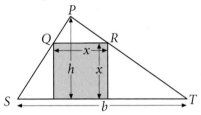

Triangles PQR and PST are similar (they share $\angle QPR$, $\angle PQR = \angle PST$ and $\angle PRQ = \angle PTS$) and the height of $\triangle PQR$ is $h - x$. Hence

$$\dfrac{b}{h} = \dfrac{x}{h-x}$$
$$\Rightarrow xh = b(h-x)$$
$$\Rightarrow xh = bh - bx$$
$$\Rightarrow xh + bx = bh$$
$$\Rightarrow x(b+h) = bh$$
$$\Rightarrow x = \dfrac{bh}{b+h} = \dfrac{1}{2}\left(\dfrac{2bh}{b+h}\right)$$

i.e. half the harmonic mean of b and h.

4 One proof is as follows. In diagram 1, a and c are parallel so the two shaded triangles are similar (equal angles are shown). A dotted line is drawn through the point of intersection that is parallel to the left-hand edge of the trapezium and lengths l_1 and l_2 are labelled on the triangles as shown.

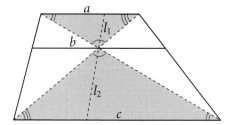

Diagram 1

Since the triangles are similar $\dfrac{l_2}{l_1} = \dfrac{c}{a}$.

In diagram 2, dotted lines are drawn as shown to be parallel to the left-hand edge of the trapezium. Their lengths are l_1 and l_2 as before. Again, the two shaded triangles are similar.

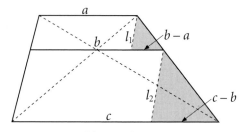

Diagram 2

The base of the smaller shaded triangle is $b - a$ and the base of the larger shaded triangle is $c - b$.

So, again by similar triangles, $\dfrac{l_2}{l_1} = \dfrac{c - b}{b - a}$.

Putting the two results together gives

$$\frac{c}{a} = \frac{c - b}{b - a}$$
$$\Rightarrow c(b - a) = a(c - b)$$
$$\Rightarrow bc - ac = ac - ab$$
$$\Rightarrow ab + bc = 2ac$$
$$\Rightarrow b(a + c) = 2ac$$
$$\Rightarrow b = \frac{2ac}{a + c}$$

which is the harmonic mean of a and c.

E Further division (p 16)

E1 $3 + \dfrac{4}{x + 2}$

E2 $x + 4 + \dfrac{5}{x - 1}$

E3 $x^2 + 3x - 1 - \dfrac{4}{x - 5}$

E4 (a) $A = 1, B = 3$ (b) $x + 1 + \dfrac{3x}{x^2 - 1}$

E5 $x + 3 + \dfrac{3x - 2}{x^2 + 1}$

Exercise E (p 18)

1 (a) $1 + \dfrac{3}{x + 2}$ (b) $1 - \dfrac{4}{x - 1}$ (c) $2 + \dfrac{3}{x^2}$

(d) $1 - \dfrac{8}{x^2 + 6}$ (e) $2 + \dfrac{1}{x + 3}$ (f) $3 - \dfrac{5}{x + 1}$

(g) $\dfrac{1}{2} + \dfrac{1}{2x}$ (h) $6 - \dfrac{2}{x^2 + 1}$

2 (a) $x + 2 + \dfrac{3}{x + 1}$ (b) $x - 4 - \dfrac{3}{x + 2}$

(c) $x - 5 + \dfrac{1}{x - 4}$ (d) $x - \dfrac{5}{x - 2}$

(e) $2x + 3 - \dfrac{1}{x - 1}$ (f) $3x - 1 + \dfrac{3}{x + 3}$

(g) $x - 2 + \dfrac{9}{2x + 1}$ (h) $2x - 5 - \dfrac{2}{3x - 2}$

3 (a) $x^2 + 2x + 3 + \dfrac{6}{x + 1}$ (b) $x^2 + 1 - \dfrac{3}{x - 5}$

(c) $x^2 - 3x - 2 + \dfrac{1}{2x - 1}$ (d) $x + 1 + \dfrac{x + 4}{x^2 + 1}$

(e) $3 + \dfrac{x - 10}{x^2 + x}$ (f) $3x - 2 - \dfrac{7}{2x^2 - 1}$

4 $\dfrac{x^3 + 1}{x^2 - 1} = x + \dfrac{x + 1}{x^2 - 1} = x + \dfrac{x + 1}{(x + 1)(x - 1)} = x + \dfrac{1}{x - 1}$

or

$\dfrac{x^3 + 1}{x^2 - 1} = \dfrac{(x + 1)(x^2 - x + 1)}{(x + 1)(x - 1)} = \dfrac{x^2 - x + 1}{x - 1} = x + \dfrac{1}{x - 1}$

5 (a) $x - 1 + \dfrac{7}{x + 1}$ (b) $x^2 + 3x + 9 + \dfrac{32}{x - 3}$

(c) $x^2 - 2x + 5 - \dfrac{10}{x + 2}$ (d) $x + \dfrac{5x}{x^2 - 5}$

Test yourself (p 19)

1 $\dfrac{5x}{x - 1}$

2 $\dfrac{x^2+5x+4}{x^2+4x}=\dfrac{(x+1)(x+4)}{x(x+4)}=\dfrac{x+1}{x}=\dfrac{x}{x}+\dfrac{1}{x}$

$=1+\dfrac{1}{x}$

3 (a) $\dfrac{x+6}{x(x+3)}$ **(b)** $\dfrac{3x}{(x+4)(x-2)}$

(c) $\dfrac{x^2+y^2}{xy}$

4 $\dfrac{6}{(y-3)(y-1)}-\dfrac{3}{y-3}$

$=\dfrac{6-3(y-1)}{(y-3)(y-1)}$

$=\dfrac{6-3y+3}{(y-3)(y-1)}$

$=\dfrac{-3y+9}{(y-3)(y-1)}$

$=\dfrac{-3(y-3)}{(y-3)(y-1)}\equiv\dfrac{-3}{y-1}\equiv\dfrac{3}{1-y}$

5 $\dfrac{4}{(y+1)(y+3)}$

6 $\dfrac{3(x+2)}{(x+1)(x-5)}$

7 $\dfrac{x-4}{x(x+1)}$

8 $A=3,\ B=-1,\ C=6$

9 One method is as follows:

$2x+\dfrac{1}{x-1}-\dfrac{5}{x^2+3x-4}$

$=2x+\dfrac{1}{x-1}-\dfrac{5}{(x-1)(x+4)}$

$=2x+\dfrac{x+4-5}{(x-1)(x+4)}$

$=2x+\dfrac{x-1}{(x-1)(x+4)}$

$=2x+\dfrac{1}{x+4}$

$=\dfrac{2x(x+4)+1}{x+4}$

$=\dfrac{2x^2+8x+1}{x+4}$ as required.

2 Functions

A What is a function? (p 20)

A1 (a) P: $y=x^2$, Q: $x^2+y^2=25$, R: $y^2=x$, S: $y=\sqrt{x}$

     (b) $y=x^2$: $y=16$, $y^2=x$: $y=2$ and $y=-2$

         $x^2+y^2=25$: $y=3$ and $y=-3$, $y=\sqrt{x}$: $y=2$

A2 $y^2-x^2=0,\ y^4=x$

A3 No, you cannot find the square root of a negative number.

A4 (a) $A(r)=100-\pi r^2$ **(b)** 71.73

     (c) $0\le r\le 5$

A5 (a) 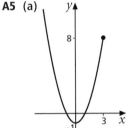   **(b)**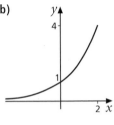

(c) 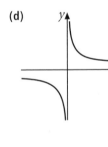   **(d)**

A6

Exercise A (p 23)

1 (a) P   **(b)** W   **(c)** V   **(d)** Q   **(e)** U

     (f) S   **(g)** R   **(h)** T

2 (a)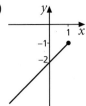

(b)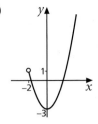

(c)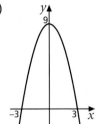

(d)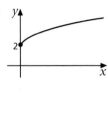

(e)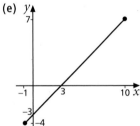

(f)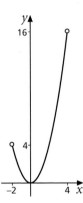

3 (a) (i) $\frac{1}{5}$ or 0.2   **(ii)** 1

    (iii) $\frac{1}{3}$ or 0.333...   **(iv)** $-\frac{1}{2}$ or -0.5

(b) -3

B Many–one and one–one functions (p 24)

B1 (a) $x = 9$ **(b)** $x = 2, -2$ **(c)** $x = 3$

B2 An explanation such as:
The line $y = 1$ cuts the graph in three places
showing that there are three different values of x
for which $f(x) = 1$. Hence g is not one–one.

B3 (a) $f(1) = 0, f(0) = 0, f(-1) = 0$ **(b)** No

B4 (a) Many–one   **(b)** One–one

    (c) One–one   **(d)** Many–one

Exercise B (p 25)

1 (a)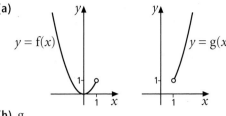

(b) g

2 (a) $t = -2, 0$

    (b) $f(t) = 3 \Rightarrow t^2 + 2t = 3$

                $\Rightarrow t^2 + 2t - 3 = 0$

                 $\Rightarrow (t + 3)(t - 1) = 0$

    If t can take any value then this equation has
two solutions $t = -3$ and $t = 1$. However, only
one of these is in the domain of f, so the
equation has only one solution, $t = 1$.

3 (a) $\sin 0 = 0, \sin \pi = 0$

    (b) π and 0 are two different values for θ in the
domain of g, both of which give the same
value for $g(\theta)$. Hence the function g is
many–one.

    (c) $\dfrac{\pi}{6} < \theta < \dfrac{5\pi}{6}$

4 h is one–one.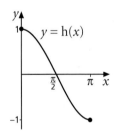

From the graph $y = h(x)$, each output can be
obtained from only one input. Hence h is
one–one.

5 (a) The area of the large square is $12^2 = 144\,\text{cm}^2$.
The area of the small square is $x^2\,\text{cm}^2$.
Hence the shaded area in cm^2 is $144 - x^2$.

    (b) $0 \le x \le 6\sqrt{2}$

    (c) (i) 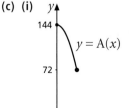   **(ii)** One–one

(d) $x = 2\sqrt{11} = 6.63$ (to 2 d.p.)

6 (a) In cm^2, the area of the base is
$(10 - 2x)(20 - 2x) = 200 - 60x + 4x^2$.
The height of the box is x cm so the volume is
$x(200 - 60x + 4x^2) = 4x^3 - 60x^2 + 200x$.

(b) $0 \le x \le 5$, $0 \le x < 5$, $0 < x \le 5$ or $0 < x < 5$

(c) One explanation is:
The expression $4x^3 - 60x^2 + 200x$ equals 0
when $x = 0$ or 5 and is positive for all values in
between. Hence the graph of $y = V(x)$ is above
the x-axis for all these values, increasing to its
maximum and decreasing back to the x-axis.
Hence there must be two values of x that lead
to each value of $V(x)$ (except for the maximum
value) so the function is many–one.

(d) We need to solve the equation
$4x^3 - 60x^2 + 200x = 144$
$\Rightarrow 4x^3 - 60x^2 + 200x - 144 = 0$
When $x = 1$, $4x^3 - 60x^2 + 200x - 144 = 0$,
so, by the factor theorem, $(x - 1)$ is a factor of
$4x^3 - 60x^2 + 200x - 144$.
Factorising gives $(x - 1)(4x^2 - 56x + 144) = 0$
$\Rightarrow 4(x - 1)(x^2 - 14x + 36) = 0$. The solutions
to $x^2 - 14x + 36 = 0$ are $x = 3.39$ and
$x = 10.61$ (to 2 d.p.). 10.61 is not in the
domain of V so the only solutions in the
domain of V are $x = 1$ and $x = 3.39$ (to 2 d.p.).
Both values will give the manufacturer a box
with the required volume though $x = 1$ will
give a very shallow box.

C The range of a function

Exercise C (p 27)

1 (a)
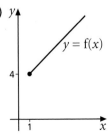
Range: $f(x) \ge 4$

(b)
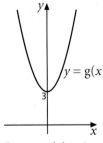
Range: $g(x) \ge 3$

(c)
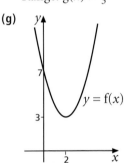
Range: $h(x) \le 1$

(d)
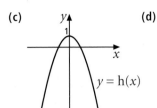
Range: $-1 \le f(x) \le 1$

(e)
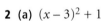
$y = g(x)$
Range: $g(x) > \frac{1}{3}$

(f)
$y = h(x)$
Range: $0 < h(x) \le 4$

(g)
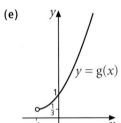
$y = f(x)$
Range: $f(x) \ge 3$

(h)
$y = g(x)$
Range: $g(x) \ge -1$

2 (a) $(x - 3)^2 + 1$

(b) (i)
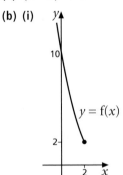
$y = f(x)$

(ii) $f(x) \ge 2$

3 (a)
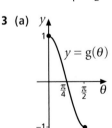
$y = g(\theta)$

(b) $-1 \le g(\theta) \le 1$

4 (a) $f(x) \geq 5$

(b) 3 is not in the range so the equation has no solution.

5 (a) One way to show that h is many–one is to find two different input values that give the same output. For example, $h(-1) = 0$ and $h(5) = 0$ so the function is many–one.

(b) $-9 \leq h(x) \leq 7$

6 (a) (i) $x = 3$ **(ii)** $x = 1$ **(iii)** $x = \frac{1}{2}$ **(iv)** $x = -1$

(b) For $h(x)$ to be 2, it would have to be the case that $\frac{1}{x} = 0$ and this is not possible.

(c) $h(x) \in \mathbb{R}, h(x) \neq 2$

7 (a) $f(x) \in \mathbb{R}, f(x) \neq -5$

(b) $f(x) \in \mathbb{R}, f(x) \neq 3$

8

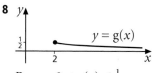

Range: $0 < g(x) \leq \frac{1}{2}$

9 $0 < f(t) < 4$

10 (a) $x = \frac{1}{4}$ **(b)** $0 < g(x) \leq 3$

11 $h(x) < 3$

D Composite functions (p 29)

D1 (a) 517 **(b)** £55

(c) (i) £56.36 **(ii)** £180.44 **(iii)** £428.60

D2 97.72

D3 (a) $gf(x) = 0.4136x + 15$ **(b)** £242.48

D4 (a) (i) 6 **(ii)** −16 **(iii)** −26

(b) (i) $32 - 3x^2$

(ii) $10 - (3x + 2)^2$ or $6 - 12x - 9x^2$

(iii) $9x + 8$

Exercise D (p 31)

1 (a) (i) 49 **(ii)** 13 **(iii)** 22

(b) $3x^2 + 1$

(c) $fg(x) = (3x + 1)^2 = 9x^2 + 6x + 1$

(d) (i) $9x + 4$ **(ii)** −5 **(iii)** $x = 5$

2 (a) $fg(x) = 2x^3 + 3, \ gf(x) = (2x + 3)^3$

(b) $fg(x) = \left(\frac{1}{x+1}\right)^2, \ gf(x) = \frac{1}{x^2 + 1}$

(c) $fg(x) = 17 - 3x, \ gf(x) = 3 - 3x$

(d) $fg(x) = 1 - (1 - 2x)^2$ or $4x - 4x^2$ or $4x(1 - x)$, $gf(x) = 2x^2 - 1$

3 $\frac{3}{2}$ or $1\frac{1}{2}$

4 $fg(x) = 3x - 1, x > 1$

5 (a) $gh(x) = g(h(x)) = g(x + 3)$
$$= 2 - \frac{6}{x+3} = \frac{2(x+3) - 6}{x+3} = \frac{2x}{x+3}$$

(b) $x = 3$

6 (a) −9

(b) gf cannot be formed as there are negative values in the range of f for which the square roots cannot be found.

7 (a) $x = 2$ **(b)** $x > 1$

8 (a) $\frac{1}{3}$ or 33.3% (to the nearest 0.1%)

(b) 20 °C

(c) $\frac{1}{3} \leq P(c) \leq \frac{3}{5}$

(d) 45.5% (to the nearest 0.1%)

(e) $Pf(t) = 1 - \frac{10}{\frac{5}{9}(t - 32)}$ or $1 - \frac{18}{t - 32}$ or $\frac{t - 50}{t - 32}$

(f) 72 °F

(g) $59 \leq t \leq 77$

E Inverse functions (p 32)

E1 (a) $gf(5) = 5, \ fg(5) = 5$

(b) $gf(-3) = -3, \ fg(-3) = -3$

(c) $gf(x) = x$

E2 (a) (i) $f(7) = 3 \times 7 - 1 = 21 - 1 = 20$

(ii) 7

(b) C: $f^{-1}(x) = \frac{x + 1}{3}$

E3 (a) $\frac{x - 3}{4}$ or $\frac{1}{4}(x - 3)$

(b) $5(x - 3)$ or $5x - 15$

(c) $\frac{1}{2}x + 7$ or $\frac{1}{2}(x + 14)$

E4 (a) $10 - x$ **(b)** $-x$ **(c)** $\frac{1}{x}$

E5 A: $f(x) = x^2 - 5, \ -3 \leq x \leq 3$

E6 (a) (i) 2   **(ii)** 5   **(iii)** 6   **(iv)** −2

(b) 10 is not in the range of g so g⁻¹(10) cannot be found.

(c) $-1 \leq x \leq 4$ or $-1 \leq y \leq 4$

(d) $-2 \leq g^{-1}(x) \leq 6$ or $-2 \leq g^{-1}(y) \leq 6$

(e)

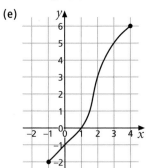

E7 (a) (i), (iv)

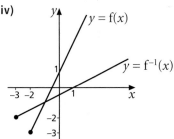

(ii) $\frac{1}{2}(x-1)$ or $\frac{1}{2}x - \frac{1}{2}$ **(iii)** $x \geq -3$

(b) (i), (iv)

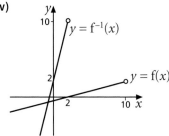

(ii) $4x + 2$ or $4(x + \frac{1}{2})$ **(iii)** $x < 2$

(c) (i), (iv)

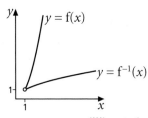

(ii) $\sqrt{x}$ **(iii)** $x > 1$

(d) (i), (iv)

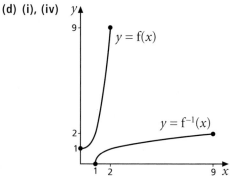

(ii) $(x-1)^{\frac{1}{3}}$ or $= \sqrt[3]{x-1}$ **(iii)** $1 \leq x \leq 9$

Exercise E (p 35)

1 $\frac{1}{5}(x-1)$ or $\frac{1}{5}x - \frac{1}{5}$

2 (a) $4(x+3)$ or $4x + 12$

(b) Domain: $-3 \leq x \leq 1$, range: $0 \leq g^{-1}(x) \leq 16$

3 (a) $\frac{9}{5}t + 32$ **(b)** $t \geq -273$ **(c)** $-94\,°F$

4 f has no inverse as it is not one–one (for example $f(1) = f(-1)$)

5 (a) $\frac{1}{3}(2-x)$ or $\frac{2}{3} - \frac{1}{3}x$ **(b)** $x < 2$ **(c)** $x = \frac{1}{2}$

6 (a) $\sqrt{x+5} + 2$ **(b)** $x > -5$

7 (a) $(x+3)^2 + 1$ **(b)** $\sqrt{x-1} - 3$

8 (a) $(x-2)^2$, $x \geq 2$

(b) $(x+5)^{\frac{1}{3}}$ or $\sqrt[3]{x+5}$, $x \in \mathbb{R}$

(c) $\dfrac{2-3x}{x}$ or $\dfrac{2}{x} - 3$, $0 < x < \frac{2}{3}$

(d) $\dfrac{5}{x+4}$, $x < -4$

9 (a) $a = \frac{2}{3}, b = 1$ **(b)**

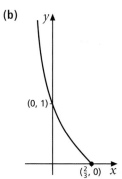

(c) $-\frac{1}{3} < x \leq \frac{2}{3}$ **(d)** $f^{-1}(x) \geq 0$

10 $y = \dfrac{2x+3}{x-2} \Rightarrow y(x-2) = 2x+3$ where $x \neq 2$

Expanding brackets gives

$yx - 2y = 2x + 3 \Rightarrow yx - 2x = 2y + 3 \Rightarrow$

$x(y-2) = 2y+3 \Rightarrow x = \dfrac{2y+3}{y-2}$, where $y \neq 2$

So $f^{-1}(x) = \dfrac{2x+3}{x-2} = f(x)$ for $x \in \mathbb{R}, x \neq 2$

11 (a) $g^{-1}(x) = \sqrt{x-4} + 1$

(b)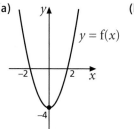

(c) Domain: $x > 4$

Range: $g^{-1}(x) > 1$

(d) $x = 1 + \sqrt{3}$

(e) The graphs of $y = g(x)$ and $y = g^{-1}(x)$ do not intersect so $g^{-1}(x) = g(x)$ has no real solutions.

Mixed questions (p 37)

1 (a)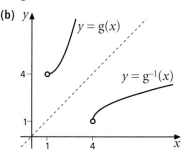

(b) $f(x) \geq -4$

(c) The function f is not one–one, for example $f(2) = f(-2) = 0$, so two different inputs lead to the same output.

2 (a) $f(x) \geq -2$

(b) -3 is not in the range of f

(c) (i) $x \geq -2$ **(ii)** $\sqrt{x+3} - 1$

3 $0 < f(t) \leq 25$

4 (a) $f(x) = \dfrac{2(2x+1)-6}{(x-1)(2x+1)}$

$= \dfrac{4x-4}{(x-1)(2x+1)}$

$= \dfrac{4(x-1)}{(x-1)(2x+1)}$

$= \dfrac{4}{2x+1}$ as required

(b) $0 < f(x) < \frac{4}{3}$

(c) $\dfrac{4-x}{2x}$ or $\dfrac{2}{x} - \dfrac{1}{2}$

(d) $f^{-1}(x) > 1$

5 (a) (i) $gf(x) = g(f(x))$

$= g\left(\dfrac{x}{x-3}\right)$

$= \dfrac{1}{\left(\dfrac{2x}{x-3} - 1\right)}$

$= \dfrac{1}{\left(\dfrac{2x-(x-3)}{x-3}\right)}$

$= \dfrac{1}{\left(\dfrac{x+3}{x-3}\right)}$

$= \dfrac{x-3}{x+3}$

$= \dfrac{x+3-6}{x+3} = 1 - \dfrac{6}{x+3}$

(ii) $x = -4$

(b) (i) $\dfrac{3x}{x-1}$

(ii) $x \in \mathbb{R}, x \neq 1$

6 (a) $(x-3)^2 + 1$

(b) k is a one–one function. As $y = k^{-1}(x)$ is the reflection of $y = k(x)$ in the line $y = x$, any points of intersection must lie on the line $y = x$ and so must satisfy $k^{-1}(x) = x$. Hence $x = 5$.

1 (a) $27x^3$

(b) (i) $a = -1, b = 1$

(ii)

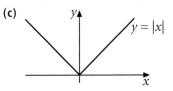

(c) $\sqrt[3]{x} + 1$

2 Graph C does not represent a function as each value of x such that $-1 < x < 1$ generates two different y-values

3 (a) $x \geq 0$ **(b)** $x = 0, \frac{1}{2}$

4 (a) $h(x) \leq 8$

(b) $x = \pm 1$

(c) The function h is not one–one, for example $h(1) = h(-1) = 5$ so two different inputs lead to the same output.

5 (a) (i) $\dfrac{3 + x}{2}$ **(ii)** $x > 0$

(b) $0 < f(x) < \frac{10}{3}$

(c) (i) $A = 10, B = -3$ **(ii)** $x = 2$

6 (a) $f(x) \geq 3$ **(b)** 2   **(c)** $x = 5, -1$

3 The modulus function

A Introducing the modulus function

Exercise A (p 40)

1 (a) 4   **(b)** 2   **(c)** 6   **(d)** 6

2 (a) 1   **(b)** 1   **(c)** 3   **(d)** 9   **(e)** 9

3 (a) 2   **(b)** 3   **(c)** 6   **(d)** 7   **(e)** $2\frac{1}{2}$

4 (a) 4   **(b)** 4   **(c)** 5   **(d)** 1   **(e)** 11

5 (a) 2   **(b)** 4   **(c)** 1   **(d)** 6   **(e)** 0

B Graphs (p 41)

B1 (a) The values in the second row of the table are 3, 2, 1, 0, 1, 2, 3.

(b) (i) $\frac{1}{2}$ **(ii)** $\frac{3}{4}$ **(iii)** 2.25

(c)

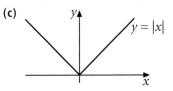

B2 (a) The values in the second row of the table are 2, 1, 0, 1, 2, 3, 4.

(b)

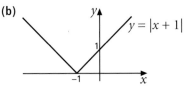

B3 (a)

x	-3	-2	-1	0	1	2	3
$\left\lvert x^2 - 4 \right\rvert$	5	0	3	4	3	0	5

(b)

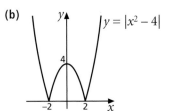

B4 (a) C   **(b)** B   **(c)** F   **(d)** E   **(e)** A   **(f)** D

B5 One explanation is that $2 - x = -(x - 2)$, so $\lvert 2 - x \rvert = \lvert -(x - 2) \rvert = \lvert x - 2 \rvert$ and so the graphs are the same.

1 (a)

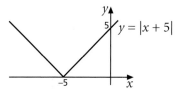
$y = |x + 5|$

(b)

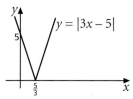
$y = |3x - 5|$

(c)

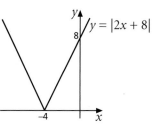
$y = |2x + 8|$

2 (a)

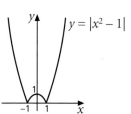
$y = |x^2 - 1|$

(b)

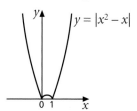
$y = |x^2 - x|$

(c)

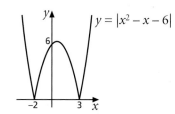
$y = |x^2 - x - 6|$

3 One explanation is that $x^2 - 9 = -(9 - x^2)$, so $|x^2 - 9| = |-(9 - x^2)| = |9 - x^2|$ and so the graphs are the same.

4

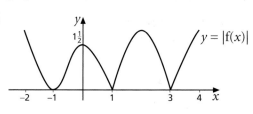
$y = |f(x)|$

5

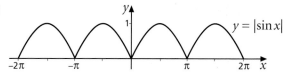
$y = |\sin x|$

6

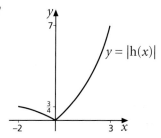
$y = |g(x)|$

7

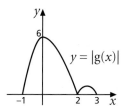
$y = |h(x)|$

8

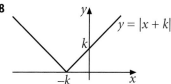
$y = |x + k|$

9 (a) $y = |x - 3|$ (b) $y = |2x - 6|$ (c) $y = |4x + 4|$

C Equations and inequalities (p 44)

C1 $|-4 + 1| = |-3| = 3$
Another solution is $x = 2$.

C2 (a) $x = 1, -9$ (b) $x = 6, -4$ (c) $x = -2, -4$
(d) $x = -6$

C3 $|x + 2|$ is greater than or equal to 0 for all values of x. So $|x + 2|$ is never negative.

C4 (a) $x = 1, -4$ (b) $x = \frac{1}{3}, -1$ (c) $x = \frac{1}{2}, 4\frac{1}{2}$

C5 There is no solution as can be seen by drawing the graphs of $y = |3x + 2|$ and $y = x$ and observing that they do not intersect.

C6 (a)

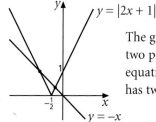
$y = |2x + 1|$
$y = -x$

The graphs intersect at two points, showing the equation $|2x + 1| = -x$ has two solutions.

(b) $x = -1, -\frac{1}{3}$

C7 $x = \frac{4}{3}, 6$

C8 (a)

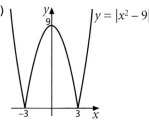

(b) (i) 4   **(ii)** 0   **(iii)** 2

(c) $x = \pm 4, \pm\sqrt{2}$

C9 (a)

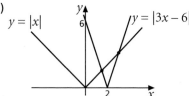

(b) $x = 3, 1\frac{1}{2}$ **(c)** $1\frac{1}{2} < x < 3$

C10

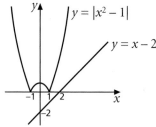

There is no solution to $|x^2 - 1| = x - 2$ as can be seen by drawing the graphs of $y = |x^2 - 1|$ and $y = x - 2$ and observing that they do not intersect.

C11 $x \le -1$ and $x \ge \frac{1}{2}$

C12 $x = \pm 3$

Exercise C (p 47)

1 (a) $x = 3, -15$ **(b)** $x = 5\frac{1}{2}, -4\frac{1}{2}$ **(c)** $x = 4, -4$

(d) $x = \pm\sqrt{3}, \pm 1$ **(e)** $x = \pm\dfrac{3}{\sqrt{2}}$ **(f)** $x = -1, 3$

2 (a) $3 < x < 7$ **(b)** $x \le -\frac{5}{3}$ and $x \ge -1$

3 (a)

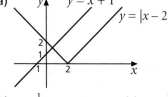

(b) $x = \frac{1}{2}$ **(c)** $x > \frac{1}{2}$

4 (a) $x = 3, \frac{1}{5}$ **(b)** $x = -4, 2$

(c) $x = -1, 2$ **(d)** $x = -1, 0, 1$

(e) $x = -2, 3$ **(f)** $x = 4, -1 - \sqrt{11}$

5 (a) The diagram below shows the graphs of $y = |x + 4|$ and $y = |x - 1|$.

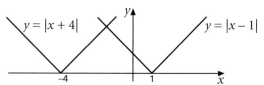

The left-hand segments are parallel as are the right-hand segments. Hence there is only the one solution to $|x + 4| = |x - 1|$.

(b) $x \ge -1\frac{1}{2}$

6 (a) $x = -\frac{2}{3}, 4$ **(b)** $x = 1\frac{1}{2}$ **(c)** $x = -2, -1$

7 (a) $x \ge -\frac{1}{2}$ **(b)** $-2 < x < 2$

(c) $x > 9$ **(d)** $x \le 2\frac{2}{3}, x \ge 12$

(e) $-3 < x < 4$ **(f)** $x \le -3, x \ge 2$

8 (a)

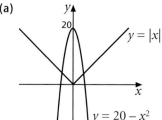

(b) $x = \pm 4$

9 (a)

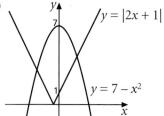

(b) $-2 < x < \sqrt{7} - 1$

10 $-2 - \sqrt{6} < x < 1 - \sqrt{11}, -2 + \sqrt{6} < x < 1 + \sqrt{11}$

D Further graphs and equations (p 48)

D1 (a) $f(4) = 7, f(-1) = 4$

(b) B

(c) $f(x) \ge 3$

(d) An explanation such as: '1 is not in the range of f' or 'f(x) = 1 $\Rightarrow$ $|x|$ = −2 which is not possible'.

(e) $x = -2, 2$

D2 (a) A   **(b)** $x = -3, 5$

D3

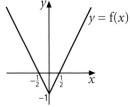

D4 (a)

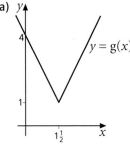

(b) An explanation such as '0 is not in the range of g' or 'g(x) = 0 $\Rightarrow$ $|2x - 3|$ = −1 which is not possible'.

(c) $x = -2, 5$

D5 (a) $f(1) = -3, f(-2) = 0$ **(b)** C

(c) $f(x) \leq 0$

D6

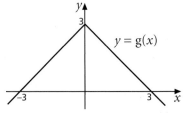

D7 (a)

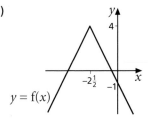

(b) $f(x) \leq 4$

D8 (a) $fg(x) = 6 - |x - 2|$, $gf(x) = |4 - x|$

(b)

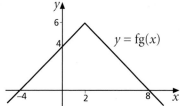

Exercise D (p 51)

1 (a)

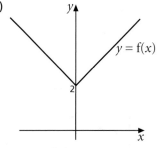

Range: $f(x) \geq 2$

(b)

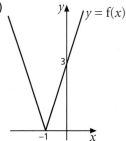

Range: $f(x) \geq 0$

(c)

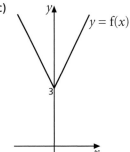

Range: $f(x) \geq 3$

(d)

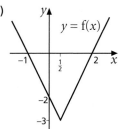

Range: $f(x) \geq -3$

(e)

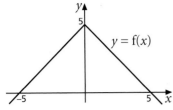

Range: f(x) ≤ 5

(f)

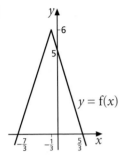

Range: f(x) ≤ 6

2 (a) $x = -6, 10$ **(b)** $x = -1, 3$ **(c)** $x = 2$

3 (a) $x = -6, 1$

  (b) An explanation such as '9 is not in the range of f' or 'f(x) = 9 $\Rightarrow$ |2x + 5| = -1 which is not possible'.

  (c) $x \leq -5\frac{1}{2}, x \geq \frac{1}{2}$

4 (a) $x = 2, 3$

  (b) The diagram below shows the graphs of $y = 1 + |3x - 7|$ and $y = x - 2$. There are no points of intersection and hence no solutions to the equation $1 + |3x - 7| = x - 2$.

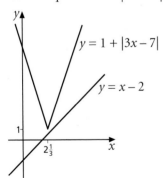

5 $x < -\frac{1}{2}, x > 1$

6 $x < 3$

7 (a)

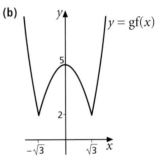

  (b) There is only one point of intersection and hence, only one solution to the equation $|3x + 2| = \frac{1}{x}$.

  Multiplying each side by x gives us $x|3x + 2| = 1$ and hence there is only one solution to this equation too.

  The solution is $x = \frac{1}{3}$.

8 f is not one–one because, for example, f(3) = f(−3) so f has no inverse.

9 (a) $fg(x) = |(x + 2)^2 - 3| = |x^2 + 4x + 1|$
  $gf(x) = |x^2 - 3| + 2$

  (b)

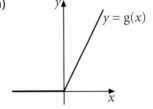

10 (a) $x = -9, 7$ **(b)** $x > 1$

11 (a)

  (b) $x = 5$

12 $x = -2, 2, \frac{1}{2}(\sqrt{17} - 1), \frac{1}{2}(1 - \sqrt{17})$

E Graphing $y = f(|x|)$ (p 52)

E1 (a) 4   (b) 4

(c) The second row of values is:
4, 3, 2, 1, 0, 1, 2, 3, 4
The third row of values is:
4, 0, −2, −2, 0, −2, −2, 0, 4

(d) (i) The values are the same.

(ii) The y-axis is a line of symmetry.

(iii)

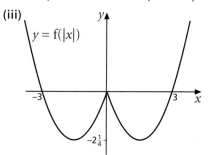

Exercise E (p 53)

1 (a) (i)

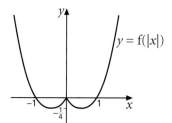

(ii)

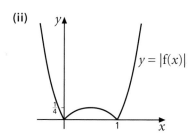

(b) $x \geq 1$ and $x = 0$

2

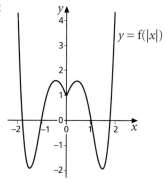

3 (a)

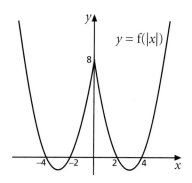

(b) (i) $x = -5, -1, 1, 5$ (ii) $x = -6, 0, 6$

(iii) $x = -8, 8$

4 (a)

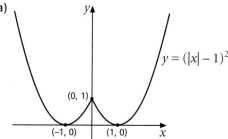

(b)

(c)

(d)

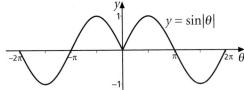

$y = \sin|\theta|$

Test yourself (p 55)

1 (a)

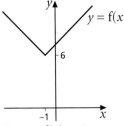

$y = f(x)$

Range: $f(x) \geq 6$

(b) 17   **(c)** $x = -5, 3$

2 (a)

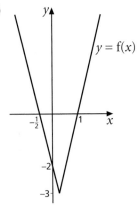

$y = f(x)$

(b) $f(x) \geq -3$ **(c)** $x = -\frac{1}{3}, 2$

3 (a) (i)

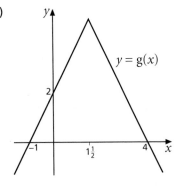

$y = g(x)$

    (ii) $g(x) \leq 5$

(b) (i) $k > 5$ **(ii)** $x = 0, 3$

(c) $-2 < x < \frac{8}{3}$

4 $x = \frac{2}{3}$

5 (a)

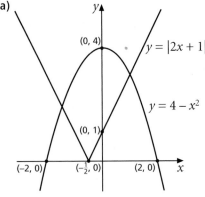

$y = |2x + 1|$

$y = 4 - x^2$

(b) $x = 1 - \sqrt{6}, 1$

6 $-4 < x < 0$

7 (a), (b)

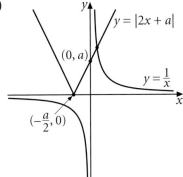

$y = |2x + a|$

$(0, a)$

$y = \frac{1}{x}$

$\left(-\frac{a}{2}, 0\right)$

(c) Where the graphs intersect we have $|2x + a| = \frac{1}{x}$ and rearranging gives $x|2x + a| = 1 \Rightarrow x|2x + a| - 1 = 0$. Since there is only one point of intersection, there is only one solution.

(d) $x = \frac{1}{2}$

8 $x = \frac{1}{2}, -\frac{1}{2}$

4 Transforming graphs

A Single transformations: revision (p 56)

A1 For a stretch of scale factor $\frac{1}{2}$ in the x-direction, replace x by $2x$ to obtain $y = (2x)^2 + 1 = 4x^2 + 1$.

A2 Let (x_1, y_1) be a point on $y = f(x)$ and let (x_2, y_2) be its image on the translated curve.
Then $(x_2, y_2) = (x_1 + a, y_1)$, giving $x_1 = x_2 - a$ and $y_1 = y_2$. We know that $y_1 = f(x_1)$, so it must be true that $y_2 = f(x_2 - a)$ and so $y = f(x - a)$ is the equation of the image.

A3 $y = |x - 3|$

A4 $y = f(x) + a$

A5 $y = 3^x + 5$

A6 (a) $y = -f(x)$ (b) $y = f(-x)$

A7 $y = x^2 - x$

Exercise A (p 59)

1 (a) A stretch of factor 3 in the y-direction
  (b) A stretch of factor $\frac{1}{3}$ in the x-direction
  (c) A stretch of factor $\frac{1}{3}$ in the y-direction
  (d) A stretch of factor 3 in the x-direction

2 $y = 2x^3$

3 (a) $f(5x) = (5x)^2 - (5x) = 25x^2 - 5x$
  (b) A stretch of factor $\frac{1}{5}$ in the x-direction
  (c)

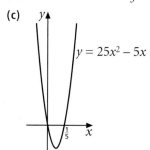

$y = 25x^2 - 5x$

4 (a) A translation of $\begin{bmatrix} 0 \\ 3 \end{bmatrix}$ (b) A translation of $\begin{bmatrix} -3 \\ 0 \end{bmatrix}$
  (c) A translation of $\begin{bmatrix} 0 \\ -3 \end{bmatrix}$ (d) A translation of $\begin{bmatrix} 3 \\ 0 \end{bmatrix}$

5 A translation of $\begin{bmatrix} -90 \\ 0 \end{bmatrix}$

6 (a) $f(x - 4) = (x - 4)^2 + (x - 4)$
    $= x^2 - 8x + 16 + x - 4 = x^2 - 7x + 12$
  (b) A translation of $\begin{bmatrix} 4 \\ 0 \end{bmatrix}$

7 $y = x^3 - 3x^2 - x + 3$

8 (a) $y = (-x)^4 + (-x)^2 = x^4 + x^2$
  (b) The equation remains the same so the y-axis must be a line of symmetry.

9 (a) (i) $y = 5^{x+1}$ (ii) $y = 5 \times 5^x$
  (b) $5 \times 5^x = 5^1 \times 5^x = 5^{x+1}$ so the images are the same.

B Combining transformations (p 60)

B1

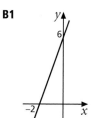

The equation is $y = 3x + 6$.

B2 (a) $y = -x - 1$ (b) $y = -x + 1$

B3 The stretch transforms $y = x^2$ to $\frac{1}{3}y = x^2$ or $y = 3x^2$. The translation transforms $y = 3x^2$ to $y + 5 = 3x^2$ which is $y = 3x^2 - 5$.

B4 (a) $y = -x^2 - 1$ (b) $y = 2(x - 5)^2$
  (c) $y = (x + 3)^2 - 2$ (d) $y = -6x^2$

B5 (a) A translation of $\begin{bmatrix} -1 \\ 0 \end{bmatrix}$ and a stretch of factor $\frac{1}{2}$ in the y-direction (the order doesn't matter)
  (b) A reflection in the x-axis followed by a translation of $\begin{bmatrix} 0 \\ 7 \end{bmatrix}$ (or a translation of $\begin{bmatrix} 0 \\ -7 \end{bmatrix}$ followed by a reflection in the x-axis)

B6 B: $y = 3f(x - 1)$

B7 B: $y = f\left(-\frac{1}{2}x\right)$

Exercise B (p 63)

1 $y = (x - 1)^2 + 5$ or $y = x^2 - 2x + 6$

2 $y = \frac{1}{8}x^3 - 2$

3 (a) $y = 2|x - 3|$

(b)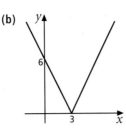

4 (a) A translation of $\begin{bmatrix} 1 \\ 0 \end{bmatrix}$ and a stretch of factor $\frac{1}{2}$ in the y-direction (the order doesn't matter)

(b) A translation of $\begin{bmatrix} 0 \\ -5 \end{bmatrix}$ and a stretch of factor $\frac{1}{3}$ in the x-direction (the order doesn't matter)

(c) A translation of $\begin{bmatrix} -1 \\ 0 \end{bmatrix}$ and a translation of $\begin{bmatrix} 0 \\ 3 \end{bmatrix}$ (the order doesn't matter)

5 (a) A stretch of factor $\frac{1}{4}$ in the x-direction and a stretch of factor 2 in the y-direction (the order doesn't matter)

(b) A translation of $\begin{bmatrix} -1 \\ 0 \end{bmatrix}$ and a stretch of factor 3 in the y-direction (the order doesn't matter)

(c) A translation of $\begin{bmatrix} 2 \\ 0 \end{bmatrix}$ and a stretch of factor $\frac{1}{4}$ in the y-direction (the order doesn't matter)

6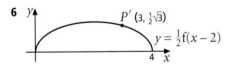

7 (a) $\frac{1}{2}f(x - 4) = \frac{1}{2}(x - 4)^2 = \frac{1}{2}(x^2 - 8x + 16)$
$= \frac{1}{2}x^2 - 4x + 8$

(b) A translation of $\begin{bmatrix} 4 \\ 0 \end{bmatrix}$ and a stretch of factor $\frac{1}{2}$ in the y-direction (the order doesn't matter)

8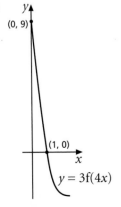

9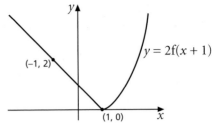

10 The image of $(0, p)$ is $(0, p + 3)$.
The image of $(q, 0)$ is $(4q, 3)$.

11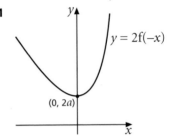

C Order of transformations (p 64)

C1 (a)

The equation is $y = 2x + 6$.

(b) $y = 2x + 3$ **(c)** No

C2 (a) $y = -(x - 1)^2$ **(b)** $y = -(x - 1)^2$

(c) An explanation such as:
The reflection replaces y by $-y$ and the translation replaces x by $(x - 1)$.
These are independent of each other so the order they are done in does not matter.

Exercise C (p 65)

1 (a) $y = 5x^2 - 10$

(b) $y = 5x^2 - 2$
The equations of the final images are different.

2 (a) $y = \frac{1}{2}f(x) - 3$ **(b)** $y = -f(x) - 4$

(c) $y = \frac{1}{4}f(x) - 1$

3 An explanation such as:
The stretch replaces y by $\frac{1}{2}y$ and the translation replaces x by $(x-3)$. These are independent of each other so the order they are done in does not matter. Either order will result in the image $y = 2f(x-3)$.

4 (a) $y = kf(x) + a$

(b) A stretch of factor 3 in the y-direction followed by a translation of $\begin{bmatrix} 0 \\ -5 \end{bmatrix}$

5 (a) A stretch of factor $\frac{1}{3}$ in the y-direction followed by a translation of $\begin{bmatrix} 0 \\ 1 \end{bmatrix}$ (or a translation of $\begin{bmatrix} 0 \\ 3 \end{bmatrix}$ followed by a stretch of factor $\frac{1}{3}$ in the y-direction)

(b) A reflection in the x-axis followed by a translation of $\begin{bmatrix} 0 \\ 5 \end{bmatrix}$ (or a translation of $\begin{bmatrix} 0 \\ -5 \end{bmatrix}$ followed by a reflection in the x-axis)

(c) A stretch of factor 3 in the y-direction followed by a translation of $\begin{bmatrix} 0 \\ -6 \end{bmatrix}$ (or a translation of $\begin{bmatrix} 0 \\ -2 \end{bmatrix}$ followed by a stretch of factor 3 in the y-direction)

6 (a) (i) $y = 2f(x) + 10$ **(ii)** $y = 2f(x) + 10$

(b) Since the images are the same the combinations of transformations are equivalent.

7 (a) (i) $y = -f(x) - 2$ **(ii)** $y = -f(x) - 2$

(b) Since the images are the same the combinations of transformations are equivalent.

Test yourself (p 67)

1 A stretch of factor $\frac{1}{2}$ in the y-direction and a translation of $\begin{bmatrix} -1 \\ 0 \end{bmatrix}$ (the order doesn't matter)

2 (a) A stretch of factor 4 in the y-direction and a translation of $\begin{bmatrix} 3 \\ 0 \end{bmatrix}$ (the order doesn't matter)

(b)

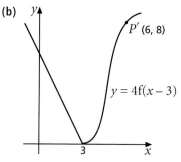

3 A stretch of factor 4 in the y-direction and a reflection in the y-axis (the order doesn't matter)

4 (a) A stretch of factor 2 in the y-direction and a stretch of factor 3 in the x-direction (the order doesn't matter)

(b)

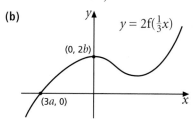

5 $y = 2x^2 - 6$

6 (a) A stretch of factor 3 in the y-direction followed by a translation of $\begin{bmatrix} 0 \\ 4 \end{bmatrix}$ (or a translation of $\begin{bmatrix} 0 \\ \frac{4}{3} \end{bmatrix}$ followed by a stretch of factor 3 in the y-direction)

(b) A reflection in the y-axis and a translation of $\begin{bmatrix} 0 \\ -7 \end{bmatrix}$ (the order doesn't matter)

7 $y = -f(x) + 6$

5 Trigonometry

A Inverse circular functions (p 68)

A1

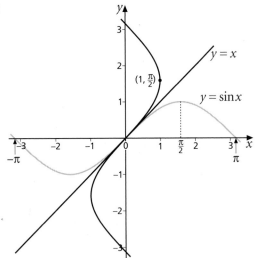

A2 No, because it is a one–many mapping, so not a function.

A3 f(x) and h(x).

A4 $-1 \leq x \leq 1$

A5 (a) 30°   (b) 45°   (c) 0°   (d) −90°

A6 (a) $\mathbb{R}$

(b), (c)

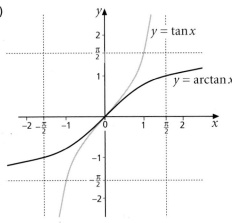

(d) Domain is $x \in \mathbb{R}$; range is $-1 \leq y \leq 1$.

Exercise A (p 70)

1 (a), (b)

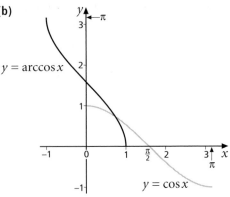

(c) Domain: $-1 \leq x \leq 1$; range: $0 \leq y \leq \pi$

2 (a) $\theta° = 60°$ (b) $\theta° = 120°$ and 240°   (c) 120°

3 (a) 180°   (b) 30°   (c) 90°   (d) −60°

4 (a) 0   (b) $\dfrac{2\pi}{3}$ (c) $-\dfrac{\pi}{4}$ (d) $-\dfrac{\pi}{3}$

  (e) $\dfrac{\pi}{2}$ (f) 0   (g) $\dfrac{3\pi}{4}$ (h) $-\dfrac{\pi}{6}$

5 (a) $x = \dfrac{1}{\sqrt{2}}$ (b) $x = \frac{1}{2}$

  (c) $x = -\dfrac{1}{\sqrt{2}}$ (d) No solution

6 (a) (i) $\frac{1}{2}$ (ii) 1   (iii) $\dfrac{\sqrt{3}}{2}$ (iv) $-\frac{1}{2}$

  (b) Yes

7 (a) (i) $\dfrac{\pi}{6}$ (ii) $\dfrac{\pi}{6}$ (iii) $-\dfrac{\pi}{6}$ (iv) $\dfrac{\pi}{6}$

  (b) No

8 (a) C   (b) $-1 \leq f(x) \leq 1$

  (c) Domain: $-1 \leq x \leq 1$
    range: $-\dfrac{\pi}{4} \leq f^{-1}(x) \leq \dfrac{\pi}{4}$

9 (a) $y = \frac{1}{2} \arcsin x$

  (b) A confirmation by graphing

B Sec, cosec and cot (p 71)

B1 (a) $\dfrac{1}{\sqrt{2}}$ (b) $\sqrt{2}$

B2 (a) $\dfrac{2}{\sqrt{3}}$ (b) 2   (c) 1   (d) $\sqrt{2}$

B3 (a) 1   (b) −1   (c) $\sqrt{2}$ (d) −2

B4 (a) 1   (b) $\sqrt{3}$ (c) $-\dfrac{1}{\sqrt{3}}$ (d) 1

B5 (a) You cannot; it is $1 \div 0$.

(b) Two other values of x which have $\sin x = 0$, such as π or 2π

(c) $x = 0, \pm\pi, \pm2\pi, \pm3\pi, \pm4\pi, \ldots$

B6 (a) Three values of x which have $\cos x = 0$, such as $\dfrac{\pi}{2}, \dfrac{3\pi}{2}, \dfrac{5\pi}{2}$

(b) $x = \pm\dfrac{\pi}{2}, \pm\dfrac{3\pi}{2}, \pm\dfrac{5\pi}{2}, \pm\dfrac{7\pi}{2}, \ldots$

B7 (a)

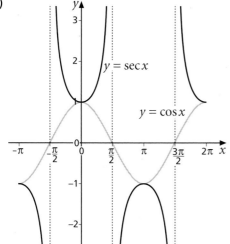

(b) $x \in \mathbb{R}, x \neq \pm\dfrac{\pi}{2}, \pm\dfrac{3\pi}{2}, \pm\dfrac{5\pi}{2}, \ldots$

(c) $y \in \mathbb{R}, y \leq -1$ and $y \geq 1$

(d) 2π

B8 (a), (b)

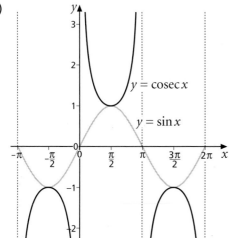

(c) Domain: $x \in \mathbb{R}, x \neq 0, \pm\pi, \pm2\pi, \pm3\pi, \ldots$
range: $x \in \mathbb{R}, x \leq -1$ and $x \geq 1$

(d) 2π

B9 $\sin^2 x + \cos^2 x = 1$

$\Rightarrow \dfrac{\sin^2 x}{\sin^2 x} + \dfrac{\cos^2 x}{\sin^2 x} = \dfrac{1}{\sin^2 x}$

$\Rightarrow 1 + \cot^2 x = \operatorname{cosec}^2 x$

Exercise B (p 73)

1 (a), (b)

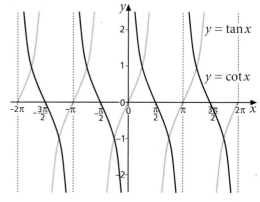

(c) Domain: $x \in \mathbb{R}, x \neq 0, \pm\pi, \pm2\pi, \pm3\pi, \ldots$
range: $x \in \mathbb{R}$

(d) π

2 (a) $\dfrac{2}{\sqrt{3}}$ **(b)** $\dfrac{1}{\sqrt{3}}$ **(c)** 2 **(d)** $-\dfrac{2}{\sqrt{3}}$

3 (a) $\sec x$ **(b)** 1 **(c)** 1 **(d)** $\tan x$

4 (a) $\cot\theta$ **(b)** $\cot\theta$ **(c)** $\operatorname{cosec} x$ **(d)** $\cot x$

5 (a) $\sin^3 x$ **(b)** 1

6 (a) $\tan\theta + \cot\theta = \tan\theta + \dfrac{1}{\tan\theta} = \dfrac{\sin\theta}{\cos\theta} + \dfrac{\cos\theta}{\sin\theta}$

$= \dfrac{\sin^2\theta + \cos^2\theta}{\cos\theta\sin\theta} = \dfrac{1}{\cos\theta\sin\theta}$

$= \dfrac{1}{\cos\theta} \times \dfrac{1}{\sin\theta} = \sec\theta\operatorname{cosec}\theta$

(b) $\cot\theta\sec\theta = \dfrac{\cos\theta}{\sin\theta} \times \dfrac{1}{\cos\theta} = \dfrac{1}{\sin\theta} = \operatorname{cosec}\theta$

7 (a) $\cot^2 x$ **(b)** $\tan^2 x$ **(c)** $-\cot x$

8 $\operatorname{cosec}^2 x = 1 + \cot^2 x$

$\Rightarrow \operatorname{cosec}^2 x - \cot^2 x = 1$

$\Rightarrow (\operatorname{cosec} x + \cot x)(\operatorname{cosec} x - \cot x) = 1$

$\Rightarrow \operatorname{cosec} x + \cot x = \dfrac{1}{\operatorname{cosec} x - \cot x}$

9 (a) $\dfrac{3}{5}$ **(b)** $\dfrac{4}{3}$ **(c)** $\dfrac{5}{3}$ **(d)** $\dfrac{5}{4}$

10 (a) $-\dfrac{5}{12}$ **(b)** $-\dfrac{12}{13}$ **(c)** $\dfrac{5}{13}$ **(d)** $\dfrac{13}{5}$

11 (a) 1

(b) $\dfrac{1}{\sqrt{2}}$ or $-\dfrac{1}{\sqrt{2}}$

(c) $\dfrac{1}{\sqrt{2}}$ or $-\dfrac{1}{\sqrt{2}}$

(d) $\sqrt{2}$ or $-\sqrt{2}$

C Solving equations (p 74)

C1 $\tan^2 x + \sec x = 1 \Rightarrow \dfrac{\sin^2 x}{\cos^2 x} + \dfrac{1}{\cos x} = 1$

$\Rightarrow \sin^2 x + \cos x = \cos^2 x$

$\Rightarrow (1 - \cos^2 x) + \cos x = \cos^2 x$

$\Rightarrow 2\cos^2 x - \cos x - 1 = 0$

$\Rightarrow (2\cos x + 1)(\cos x - 1) = 0$

$\Rightarrow \cos x = -\tfrac{1}{2}$ or 1

$\Rightarrow x = \dfrac{2\pi}{3}$ or $\dfrac{4\pi}{3}$ or $x = 0$ or 2π

Exercise C (p 75)

1 (a) $\theta° = 45°$ or $315°$

(b) $\theta° = 45°, 135°, 225°$ or $315°$

2 $\theta° = -162°$ or $18°$

3 (a) $x = \dfrac{\pi}{4}$ or $\dfrac{5\pi}{4}$

(b) $x = \dfrac{\pi}{2}$

(c) $x = \dfrac{\pi}{2}$ or $\dfrac{3\pi}{2}$

(d) $\dfrac{\pi}{3}, \dfrac{2\pi}{3}, \dfrac{4\pi}{3}$ or $\dfrac{5\pi}{3}$

4 (a) $x = -1.23$ or 1.23

(b) $x = -1.91, -1.23, 1.23$ or 1.91

(c) $x = 0.34$ or 2.80

(d) $x = -2.03$ or 1.11

5 (a) $\sec^2 x + \tan x = 3$

$\Rightarrow (1 + \tan^2 x) + \tan x = 3$

$\Rightarrow \tan^2 x + \tan x - 2 = 0$

(b) $(\tan x - 1)(\tan x + 2)$

(c) $x = \dfrac{\pi}{4}$ (0.79), $-\dfrac{3\pi}{4}$ (-2.36), -1.11 or 2.03

6 $\theta = -\dfrac{\pi}{2}$

7 $x = 1.2$ or 5.1

8 $x = 45.0°$ or $56.3°$

9 (a) $\theta° = 78.69°, 153.43°, 258.69°$ or $333.43°$

(b) $\theta° = 14.04°, 135.00°, 194.04°$ or $315.00°$

10 There is no solution $\left(\text{note that } x = \dfrac{\pi}{2} \text{ is not a}\right.$ solution, since neither $\sec x$ nor $\tan x$ is defined for $x = \dfrac{\pi}{2}\Big)$.

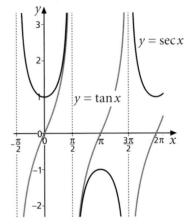

$y = \sec x$ and $y = \tan x$ do not intersect, and share the same asymptotes.

11 A $(0.90, 0.79)$ and B $(5.38, -0.79)$

D Transforming graphs (p 76)

D1 (a) A stretch in the y-direction, scale factor $\tfrac{1}{3}$; a reflection in the x- or y-axis (order unimportant)

(b) A stretch in the x-direction, scale factor $\tfrac{1}{2}$; a translation of $\begin{bmatrix} 0 \\ 3 \end{bmatrix}$ (order unimportant)

(c) A reflection in the x- or y-axis; a translation of $\begin{bmatrix} 0 \\ 3 \end{bmatrix}$ (order unimportant)

D2 (a) (i) $y = \cos\tfrac{1}{4}x + 2$ **(ii)** $y = 2\cos(x - 1)$

(iii) $y = 2\cos\tfrac{1}{3}x$

(b) (i)

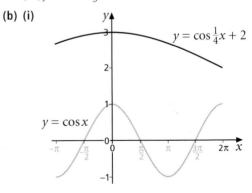

(ii)

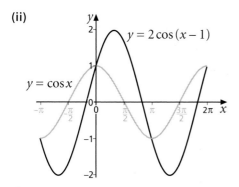

$y = 2\cos(x - 1)$

$y = \cos x$

(iii)

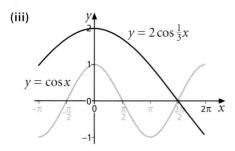

$y = 2\cos \frac{1}{3}x$

$y = \cos x$

D3 There are other possibilities besides those described here.

(a) (i) A stretch in the y-direction, scale factor $\frac{1}{2}$; then a translation of $\begin{bmatrix} \frac{\pi}{4} \\ 0 \end{bmatrix}$ (or vice versa)

(ii) $y = \frac{1}{2}\sin\left(x - \frac{\pi}{4}\right)$

(b) (i) A reflection in the x-axis; then a translation of $\begin{bmatrix} -\frac{\pi}{6} \\ 0 \end{bmatrix}$ (or vice versa)

(ii) $y = -\sin\left(x + \frac{\pi}{6}\right)$

(c) (i) A stretch in the y-direction, scale factor 2; then a stretch in the x-direction, scale factor $\frac{1}{3}$ (or vice versa)

(ii) $y = 2\sin 3x$

(d) (i) A stretch in the x-direction, scale factor 2; then a translation of $\begin{bmatrix} 0 \\ 1 \end{bmatrix}$

(ii) $y = \left(\sin \frac{1}{2}x\right) + 1$

Exercise D (p 79)

Where transformations have to be identified from given graphs there are alternative answers to those provided here.

1 (a) A stretch in the x-direction, scale factor $\frac{1}{4}$; $y = \cos 4x$

(b) A translation of $\begin{bmatrix} 0 \\ -1 \end{bmatrix}$; $y = (\cos x) - 1$

(c) A translation of $\begin{bmatrix} \frac{\pi}{2} \\ 0 \end{bmatrix}$; $y = \cos\left(x - \frac{\pi}{2}\right)$

(d) A stretch in the x-direction, scale factor 4; $y = \cos \frac{1}{4}x$

2 (a) (i) $y = \left(\tan \frac{1}{2}x\right) - 1$ (ii) $y = \tan(-2x)$

(iii) $y = -\tan(x - 1)$ (iv) $y = \tan(-x) + 1$

(b) (i)

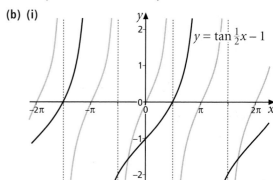

$y = \tan \frac{1}{2}x - 1$

(ii)

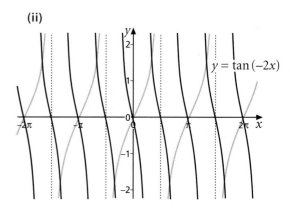

$y = \tan(-2x)$

(iii)

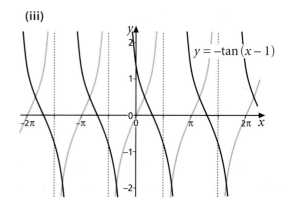

$y = -\tan(x - 1)$

(iv)

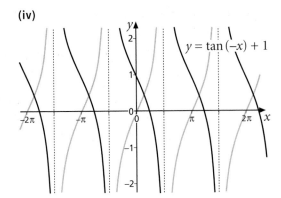

$y = \tan(-x) + 1$

(b)

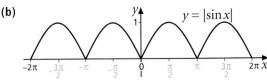

$y = |\sin x|$

(c)

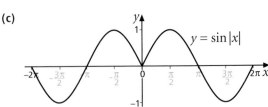

$y = \sin|x|$

3 (a) $\left(\dfrac{5\pi}{12}, 1\right)$

(b) $y = 2\sin\left(x - \dfrac{\pi}{4}\right)$

(c) When $x = \dfrac{5\pi}{12}$, $2\sin\left(x - \dfrac{\pi}{4}\right) = 2\sin\left(\dfrac{5\pi}{12} - \dfrac{\pi}{4}\right)$

$= 2\sin\left(\dfrac{\pi}{6}\right) = 1$; the image of A fits.

4 (a) (i) A stretch in the x-direction, scale factor $\frac{1}{2}$, then a stretch in the y-direction scale factor $\frac{1}{2}$ (or vice versa).

(ii) $y = \frac{1}{2}\cos 2x$

(b) (i) A stretch in the x-direction, scale factor 2, then a translation by $\begin{bmatrix} 0 \\ -1 \end{bmatrix}$ (or vice versa)

(ii) $y = \left(\cos\frac{1}{2}x\right) - 1$

(c) (i) A stretch in the y-direction, scale factor $1\frac{1}{2}$, then a translation by $\begin{bmatrix} -\dfrac{\pi}{3} \\ 0 \end{bmatrix}$ (or vice versa)

(ii) $y = 1.5\cos\left(x + \dfrac{\pi}{3}\right)$

(d) (i) A translation by $\begin{bmatrix} 0.5 \\ 0 \end{bmatrix}$ then a translation by $\begin{bmatrix} 0 \\ -0.25 \end{bmatrix}$ (or vice versa).

(ii) $y = \cos(x - 0.5) - 0.25$

5 (a)

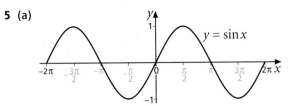

$y = \sin x$

6 (a), (b)

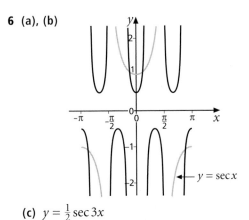

$y = \sec x$

(c) $y = \frac{1}{2}\sec 3x$

7 (a) Maxima $\left(-\dfrac{2\pi}{3}, 1\right)$, $(0, 1)$, $\left(\dfrac{2\pi}{3}, 1\right)$;

minima $(-\pi, -1)$, $\left(-\dfrac{\pi}{3}, -1\right)$, $\left(\dfrac{\pi}{3}, -1\right)$, $(\pi, -1)$

(b)

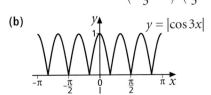

$y = |\cos 3x|$

(c) $x = -\pi, -\dfrac{2\pi}{3}, -\dfrac{\pi}{3}, 0, \dfrac{\pi}{3}, \dfrac{2\pi}{3}$ and π

8 (a) $y = 2\arcsin(x - 1)$

(b)

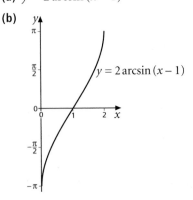

$y = 2\arcsin(x - 1)$

9 (a) $-1 \leq f(x) \leq 1$

(b) Domain: $-1 \leq x \leq 1$; range: $-\pi \leq f^{-1}(x) \leq \pi$

(c) $y = 2\arcsin x$

(d) $fg(x) = \sin\frac{1}{2}|x|$

(e)

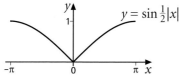

10 (a)

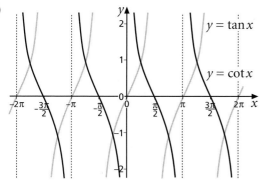

(b) Transform $y = \cot x$ by, for example, a reflection in the y-axis, followed by a translation of $\begin{bmatrix} -\dfrac{\pi}{2} \\ 0 \end{bmatrix}$

(c) $y = \cot\left(-x + \dfrac{\pi}{2}\right)$

(d) $y = \cot\left(-x + \dfrac{\pi}{2}\right) = \cot\left(\dfrac{\pi}{2} - x\right)$

$$= \dfrac{1}{\tan\left(\dfrac{\pi}{2} - x\right)} = \dfrac{1}{\left(\dfrac{\sin\left(\dfrac{\pi}{2} - x\right)}{\cos\left(\dfrac{\pi}{2} - x\right)}\right)}$$

$$= \dfrac{1}{\left(\dfrac{\cos x}{\sin x}\right)} = \dfrac{\sin x}{\cos x} = \tan x$$

Test yourself (p 83)

1 (a) $\dfrac{\pi}{6}$ **(b)** $\dfrac{\pi}{6}$ **(c)** $\dfrac{5\pi}{6}$ **(d)** $\dfrac{2\pi}{3}$

2 (a) 2 **(b)** $\sqrt{2}$ **(c)** $\dfrac{1}{\sqrt{3}}$ **(d)** 2

3 (a) $\csc x$ **(b)** $-\tan^2 x$ **(c)** $\csc^2 x$

4 $x = \dfrac{\pi}{2}, \dfrac{3\pi}{4}, \dfrac{3\pi}{2}, \dfrac{7\pi}{4}$

5 $x = 200°$ or $340°$

6 (a) A reflection in the y-axis, followed by a stretch in the y-direction, factor 2 (or vice versa)

(b) A stretch in the x-direction, factor $\frac{1}{2}$, followed by a translation by $\begin{bmatrix} 0 \\ 1 \end{bmatrix}$ (or vice versa)

(c) A reflection in the x-axis, followed by a stretch in the x-direction, factor 2 (or vice versa)

(d) A reflection in the x-axis, followed by a translation by $\begin{bmatrix} 0 \\ 2 \end{bmatrix}$

7 (a) $y = \left(\sin\frac{1}{4}x\right) + 2$ **(b)** $y = 2\sin(-x)$

(c) $y = -\sin(x + 1)$

8 (a) (i) A stretch in the x-direction, factor 2, followed by a stretch in the y-direction, factor $\frac{1}{2}$ (or vice versa)

(ii) $y = \frac{1}{2}\sin\frac{1}{2}x$

(b) (i) A stretch in the x-direction, factor $\frac{3}{2}$, followed by a stretch in the y-direction, factor $\frac{3}{2}$ (or vice versa)

(ii) $y = \frac{3}{2}\sin\frac{2}{3}x$

6 Trigonometric formulae

A Addition formulae (p 84)

A1 (a) $\cos B$

(b) $\cos B \sin A$

(c) $\angle RPO = 90° - A$
$\angle QPT = 180° - (90° + \angle RPO)$
$= 180° - (90° + 90° - A) = A$
$PT = PQ \cos \angle QPT$
In $\triangle OPQ$, $PQ = \sin B$, so $PT = \sin B \cos \angle QPT$
$= \sin B \cos A$

(d) $\cos B \sin A + \sin B \cos A$

(e) $QH = \sin(A + B) = RP + PT$
Hence $\sin(A + B) = \cos B \sin A + \sin B \cos A$
$= \sin A \cos B + \cos A \sin B$

A2 $OH = OR - HR = OR - QT$
$\cos(A + B) = OP \cos A - PQ \sin A$
$= \cos B \cos A - \sin B \sin A$
$= \cos A \cos B - \sin A \sin B$

A3 $\sin(A + B) = \sin A \cos B + \cos A \sin B$
$\Rightarrow \sin(A + (-B)) = \sin A \cos(-B) + \cos A \sin(-B)$
$\qquad\qquad = \sin A \cos B + \cos A (-\sin B)$
$\Rightarrow \quad \sin(A - B) = \sin A \cos B - \cos A \sin B$

$\cos(A + B) = \cos A \cos B - \sin A \sin B$
$\Rightarrow \cos(A + (-B)) = \cos A \cos(-B) - \sin A \sin(-B)$
$\qquad\qquad = \cos A \cos B - \sin A (-\sin B)$
$\Rightarrow \quad \cos(A + B) = \cos A \cos B + \sin A \sin B$

A4 $\sin(A + A) = \sin A \cos A + \cos A \sin A$
$\Rightarrow \sin 2A = 2 \sin A \cos A$

A5 (a) $\cos(A + A) = \cos A \cos A - \sin A \sin A$
$\Rightarrow \quad \cos 2A = \cos^2 A - \sin^2 A$
$\qquad\qquad = \cos^2 A - (1 - \cos^2 A)$
$\qquad\qquad = \cos^2 A - 1 + \cos^2 A$
$\qquad\qquad = 2 \cos^2 A - 1$

(b) $\cos 2A = \cos^2 A - \sin^2 A$
$\qquad\quad = 1 - \sin^2 A - \sin^2 A$
$\qquad\quad = 1 - 2 \sin^2 A$

A6 Given that $\sin A = \frac{1}{4}$, if A is acute, then
$0° < A < 45°$, so $0° < 2A < 90°$.
In this case $\sin 2A$ and $\cos 2A$ are both > 0.
But for $\sin A = \frac{1}{4}$, A might be obtuse, with
$135° < A < 180°$, giving $270° < 2A < 360°$.

In this case $\sin 2A < 0$ but again $\cos 2A > 0$.
Hence there is one value for $\cos 2A$ but two
for $\sin 2A$.

A7 (a) $\dfrac{\sin(A + B)}{\cos(A + B)}$

(b) $\dfrac{\sin(A + B)}{\cos(A + B)} = \dfrac{\sin A \cos B + \cos A \sin B}{\cos A \cos B - \sin A \sin B}$

(c) $\dfrac{\dfrac{\sin A \cos B}{\cos A \cos B} + \dfrac{\cos A \sin B}{\cos A \cos B}}{\dfrac{\cos A \cos B - \sin A \sin B}{\cos A \cos B}}$

$= \dfrac{\dfrac{\sin A}{\cos A} + \dfrac{\sin B}{\cos B}}{1 - \dfrac{\sin A \sin B}{\cos A \cos B}}$

$= \dfrac{\tan A + \tan B}{1 - \tan A \tan B}$

A8 (a) $\tan(A + (-B)) = \dfrac{\tan A + \tan(-B)}{1 - \tan A \tan(-B)}$

$\qquad\qquad\qquad = \dfrac{\tan A + (-\tan B)}{1 - \tan A(-\tan B)}$

$\Rightarrow \tan(A - B) = \dfrac{\tan A - \tan B}{1 + \tan A \tan B}$

(b) $\tan 2A = \tan(A + A)$
$\qquad\quad = \dfrac{\tan A + \tan A}{1 - \tan A \tan A} = \dfrac{2 \tan A}{1 - \tan^2 A}$

Exercise A (p 88)

1 $\sin(x + 180)° = \sin x° \cos 180° + \cos x° \sin 180°$
$\qquad\qquad\quad = \sin x° \times (-1) + \cos x° \times 0$
$\qquad\qquad\quad = -\sin x°$

The graph of $y = \sin(x + 180)°$ is that of $y = \sin x°$
translated by $\begin{bmatrix} -180° \\ 0 \end{bmatrix}$.

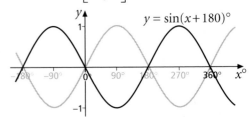

The resulting graph is the same as the reflection
of $y = \sin x°$ in the x-axis, i.e. $y = -\sin x°$.

2 (a) $\sin 75° = \sin(45° + 30°)$
$$= \sin 45° \cos 30° + \cos 45° \sin 30°$$
$$= \frac{1}{\sqrt{2}} \times \frac{\sqrt{3}}{2} + \frac{1}{\sqrt{2}} \times \frac{1}{2}$$
$$= \frac{\sqrt{3}+1}{2\sqrt{2}}$$

(b) $\sin 15° = \sin(45° - 30°)$
$$= \sin 45° \cos 30° - \cos 45° \sin 30°$$
$$= \frac{1}{\sqrt{2}} \times \frac{\sqrt{3}}{2} - \frac{1}{\sqrt{2}} \times \frac{1}{2}$$
$$= \frac{\sqrt{3}-1}{2\sqrt{2}}$$

3 (a) $\sin^2 A = 1 - \cos^2 A = 1 - \left(\frac{4}{5}\right)^2 = 1 - \frac{16}{25} = \frac{9}{25}$

So $\sin A = \pm\frac{3}{5}$

Since A is acute, $\sin A > 0$, so $\sin A = \frac{3}{5}$

(b) $\sin 2A = 2\sin A \cos A = 2 \times \frac{3}{5} \times \frac{4}{5} = \frac{24}{25}$

(c) $\cos 2A = 1 - 2\sin^2 A = 1 - 2 \times \frac{9}{25} = \frac{7}{25}$

(d) $\operatorname{cosec} 2A = \frac{1}{\sin 2A} = \frac{1}{\frac{24}{25}} = \frac{25}{24}$

4 (a) $\tan(A + B) = \frac{\tan A + \tan B}{1 - \tan A \tan B} = \frac{\frac{1}{2} + \frac{1}{3}}{1 - \frac{1}{2} \times \frac{1}{3}}$
$$= \frac{\frac{5}{6}}{1 - \frac{1}{6}} = \frac{\frac{5}{6}}{\frac{5}{6}} = 1$$

(b) $\tan 2C = \frac{3}{4} \Rightarrow \frac{2\tan C}{1 + \tan^2 C} = \frac{3}{4}$
$$\Rightarrow 4 \times 2\tan C = 3 \times (1 - \tan^2 C)$$
$$\Rightarrow 8\tan C = 3 - 3\tan^2 C$$
$$\Rightarrow 3\tan^2 C + 8\tan C - 3 = 0$$
$$\Rightarrow (3\tan C - 1)(\tan C + 3) = 0$$
$$\Rightarrow \tan C = \frac{1}{3} \text{ or } -3$$

5 (a) $\cos(A + B) + \cos(A - B)$
$$= (\cos A \cos B - \sin A \sin B)$$
$$\quad + (\cos A \cos B + \sin A \sin B)$$
$$= 2\cos A \cos B$$

(b) $\cos(A - B) - \cos(A + B)$
$$= (\cos A \cos B + \sin A \sin B)$$
$$\quad - (\cos A \cos B - \sin A \sin B)$$
$$= 2\sin A \sin B$$

6 (a) $\cos 2A \cos A - \sin 2A \sin A$
$$= \cos(2A + A) = \cos 3A$$

(b) $\cos(A + B)\cos A + \sin(A + B)\sin A$
$$= \cos((A + B) - A) = \cos B$$

(c) $2\sin 3C \cos 3C = \sin(2 \times 3C) = \sin 6C$

(d) $\sin 3D \cos 2D + \cos 3D \sin 2D$
$$= \sin(3D + 2D) = \sin 5D$$

7 (a) $(\cos A + \sin A)(\cos B + \sin B)$
$$= \cos A \cos B + \sin A \sin B$$
$$\quad + \cos A \sin B + \sin A \cos B$$
$$= \cos(A - B) + \sin(A + B)$$

(b) $(\cos A + \sin A)^2$
$$= \cos^2 A + 2\sin A \cos A + \sin^2 A$$
$$= \cos^2 A + \sin^2 A + 2\sin A \cos A$$
$$= 1 + 2\sin A \cos A = 1 + \sin 2A$$

(c) $\sin 3A = \sin(2A + A)$
$$= \sin 2A \cos A + \cos 2A \sin A$$
$$= (2\sin A \cos A) \times \cos A + (1 - 2\sin^2 A) \times \sin A$$
$$= 2\sin A \cos^2 A + \sin A - 2\sin^3 A$$
$$= 2\sin A (1 - \sin^2 A) + \sin A - 2\sin^3 A$$
$$= 2\sin A - 2\sin^3 A + \sin A - 2\sin^3 A$$
$$= 3\sin A - 4\sin^3 A$$

(d) $(\sin A + \cos B)^2 + (\cos A - \sin B)^2$
$$= (\sin^2 A + 2\sin A \cos B + \cos^2 B)$$
$$\quad + (\cos^2 A - 2\cos A \sin B + \sin^2 B)$$
$$= \sin^2 A + \cos^2 A + 2\sin A \cos B - 2\cos A \sin B$$
$$\quad + \cos^2 B + \sin^2 B$$
$$= 1 + 2(\sin A \cos B - \cos A \sin B) + 1$$
$$= 2 + 2 \times \sin(A - B) = 2(1 + \sin(A - B))$$

8 $\cot 2x + \operatorname{cosec} 2x = \frac{\cos 2x}{\sin 2x} + \frac{1}{\sin 2x}$
$$= \frac{\cos 2x + 1}{\sin 2x} = \frac{\cos 2x + 1}{2\sin x \cos x} = \frac{(2\cos^2 x - 1) + 1}{2\sin x \cos x}$$
$$= \frac{2\cos^2 x}{2\sin x \cos x} = \frac{\cos x}{\sin x} = \cot x$$

9 $\dfrac{\cos\theta}{1 - \sqrt{2}\sin\theta} - \dfrac{\cos\theta}{1 + \sqrt{2}\sin\theta}$
$$= \frac{\cos\theta(1 + \sqrt{2}\sin\theta) - \cos\theta(1 - \sqrt{2}\sin\theta)}{(1 - \sqrt{2}\sin\theta)(1 + \sqrt{2}\sin\theta)}$$
$$= \frac{\cos\theta + \sqrt{2}\cos\theta\sin\theta - \cos\theta + \sqrt{2}\cos\theta\sin\theta}{1 - 2\sin^2\theta}$$
$$= \frac{2\sqrt{2}\cos\theta\sin\theta}{1 - 2\sin^2\theta} = \frac{\sqrt{2}\sin 2\theta}{\cos 2\theta}$$
$$= \sqrt{2}\tan 2\theta$$

10 $\dfrac{\tan a}{\sec a - 1} = \dfrac{\dfrac{\sin a}{\cos a}}{\dfrac{1}{\cos a} - 1} = \dfrac{\sin a}{1 - \cos a}$

$= \dfrac{2 \sin \frac{1}{2}a \cos \frac{1}{2}a}{1 - (1 - 2 \sin^2 \frac{1}{2}a)} = \dfrac{2 \sin \frac{1}{2}a \cos \frac{1}{2}a}{2 \sin^2 \frac{1}{2}a} = \dfrac{\cos \frac{1}{2}a}{\sin \frac{1}{2}a}$

$= \cot \frac{1}{2}a$

11 (a) $\dfrac{2t}{1 + t^2} = \dfrac{2 \tan \frac{1}{2}\theta}{1 + \tan^2 \frac{1}{2}\theta} = \dfrac{\dfrac{2 \sin \frac{1}{2}\theta}{\cos \frac{1}{2}\theta}}{1 + \dfrac{\sin^2 \frac{1}{2}\theta}{\cos^2 \frac{1}{2}\theta}}$

$= \dfrac{\cos^2 \frac{1}{2}\theta \times \dfrac{2 \sin \frac{1}{2}\theta}{\cos \frac{1}{2}\theta}}{\cos^2 \frac{1}{2}\theta + \sin^2 \frac{1}{2}\theta} = \dfrac{\cos \frac{1}{2}\theta \times 2 \sin \frac{1}{2}\theta}{1}$

$= 2 \cos \frac{1}{2}\theta \sin \frac{1}{2}\theta = \sin \theta$

(b) $\dfrac{1 - t^2}{1 + t^2} = \dfrac{1 - \tan^2 \frac{1}{2}\theta}{1 + \tan^2 \frac{1}{2}\theta} = \dfrac{1 - \dfrac{\sin^2 \frac{1}{2}\theta}{\cos^2 \frac{1}{2}\theta}}{1 + \dfrac{\sin^2 \frac{1}{2}\theta}{\cos^2 \frac{1}{2}\theta}}$

$= \dfrac{\cos^2 \frac{1}{2}\theta - \sin^2 \frac{1}{2}\theta}{\cos^2 \frac{1}{2}\theta + \sin^2 \frac{1}{2}\theta} = \dfrac{\cos^2 \frac{1}{2}\theta - \sin^2 \frac{1}{2}\theta}{1}$

$= \cos \theta$

12 (a) $\sin 2x° = \cos x°$

$\Rightarrow 2 \sin x° \cos x° = \cos x°$

$\Rightarrow 2 \sin x° \cos x° - \cos x° = 0$

$\Rightarrow \cos x°(2 \sin x° - 1) = 0$

$\Rightarrow \cos x° = 0$ or $\sin x° = \frac{1}{2}$

$\cos x° = 0$ gives $x° = 90°$ or $270°$

$\sin x° = \frac{1}{2}$ gives $x° = 30°$ or $150°$

So $x° = 30°, 90°, 150°$ or $270°$

(b) $\sin x° + \cos 2x° = 0$

$\Rightarrow \sin x° + (1 - 2 \sin^2 x°) = 0$

$\Rightarrow 2 \sin^2 x° - \sin x° - 1 = 0$

$\Rightarrow (2 \sin x° + 1)(\sin x° - 1) = 0$

$\Rightarrow 2 \sin x° + 1 = 0$ or $\sin x° - 1 = 0$

$\Rightarrow \sin x° = -\frac{1}{2}$ or $\sin x° = 1$

$\sin x° = -\frac{1}{2}$ gives $x° = 210°$ or $330°$

$\sin x° = 1$ gives $x° = 90°$

So $x° = 90°, 210°$ or $330°$

(c) $\cos 2x° = 7 \cos x° + 3$

$\Rightarrow 2 \cos^2 x° - 1 = 7 \cos x° + 3$

$\Rightarrow 2 \cos^2 x° - 7 \cos x° - 4 = 0$

$\Rightarrow (2 \cos x° + 1)(\cos x° - 4) = 0$

$\Rightarrow 2 \cos x° + 1 = 0$ or $\cos x° - 4 = 0$

$\Rightarrow \cos x° = -\frac{1}{2}$ or $\cos x° = 4$ (no solution)

So $x° = 120°$ or $240°$

(d) $\cos 2x° = 1 + \sin x°$

$\Rightarrow 1 - 2 \sin^2 x° = 1 + \sin x°$

$\Rightarrow -2 \sin^2 x° = \sin x°$

$\Rightarrow 2 \sin^2 x° + \sin x° = 0$

$\Rightarrow \sin x°(2 \sin x° + 1) = 0$

$\Rightarrow \sin x° = 0$ or

$2 \sin x° + 1 = 0 \left(\text{so } \sin x° = -\frac{1}{2}\right)$

$\sin x° = 0$ gives $x° = 0°, 180°$ or $360°$

$\sin x° = -\frac{1}{2}$ gives $x° = 210°$ or $330°$

So $x° = 0°, 180°, 210°, 330°$ or $360°$

(e) $\sin 2x° = \tan x°$

$\Rightarrow 2 \sin x° \cos x° = \dfrac{\sin x°}{\cos x°}$

$\Rightarrow \cos x° \times 2 \sin x° \cos x° = \sin x°$

$\Rightarrow 2 \cos^2 x° \sin x° - \sin x° = 0$

$\Rightarrow \sin x°(2 \cos^2 x° - 1) = 0$

$\Rightarrow \sin x° = 0$ or

$2 \cos^2 x° - 1 = 0 \left(\text{so } \cos x° = \pm \dfrac{1}{\sqrt{2}}\right)$

$\sin x° = 0$ gives $x° = 0°, 180°$ or $360°$

$\cos x° = \pm \dfrac{1}{\sqrt{2}}$ gives $x° = 45°, 135°, 225°$ or $315°$

$\Rightarrow x° = 0°, 45°, 135°, 180°, 225°, 315°$ or $360°$

13 (a) $\sin(\theta + \phi) + \sin(\theta - \phi)$

$= \sin \theta \cos \phi + \cos \theta \sin \phi$

$\quad + \sin \theta \cos \phi - \cos \theta \sin \phi$

$= 2 \sin \theta \cos \phi$

(b) $\sin(\theta + \phi) - \sin(\theta - \phi)$

$= \sin \theta \cos \phi + \cos \theta \sin \phi$

$\quad - (\sin \theta \cos \phi - \cos \theta \sin \phi)$

$= \sin \theta \cos \phi + \cos \theta \sin \phi$

$\quad - \sin \theta \cos \phi + \cos \theta \sin \phi$

$= 2 \cos \theta \sin \phi$

14 (a) $\sin(A + B) = \sin A \cos B + \cos A \sin B$
$\sin(A - B) = \sin A \cos B - \cos A \sin B$
So $\sin(A + B) + \sin(A - B) = 2\sin A \cos B$
Given that $X = A + B$ and $Y = A - B$,
$X + Y = 2A \implies A = \dfrac{X + Y}{2}$
and $X - Y = 2B \implies B = \dfrac{X - Y}{2}$
Substituting these into
$\sin(A + B) + \sin(A - B) = 2\sin A \cos B$
we have
$\sin X + \sin Y = 2\sin \dfrac{X + Y}{2} \cos \dfrac{X - Y}{2}$

(b) Substituting 4θ for X and 2θ for Y in the identity proved above,
$\sin 4\theta + \sin 2\theta = 2\sin \dfrac{4\theta + 2\theta}{2} \cos \dfrac{4\theta - 2\theta}{2}$
So the equation may be written as
$2\sin 3\theta \cos \theta = 0$
$\implies \sin 3\theta = 0$ or $\cos \theta = 0$
$\sin 3\theta = 0$ gives $3\theta = 0°, 180°, 360°, 540°, 720°, 900°$ or $1080°$, from which
$\theta = 0°, 60°, 120°, 180°, 240°, 300°, 360°$
$\cos \theta = 0$ gives $\theta = 90°$ or $270°$
So $\theta = 0°, 60°, 90°, 120°, 180°, 240°, 270°, 300°,$ or $360°$

B Equivalent expressions (p 89)

B1 (a) $x° = 0°$ or $120°$ or $360°$

(b) $x° = 2°$ or $98°$

(c) $x° = 143°$ or $323°$

B2 You cannot solve it using methods used in B1. Dividing by (say) $\cos x°$ gives an equation involving $\tan x°$ and $\operatorname{cosec} x°$.

B3 (a)

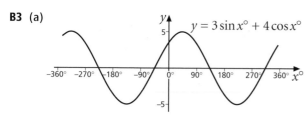

(b) (i) Period $= 360°$ **(ii)** Amplitude $= 5$

B4 (a) Period $= 360°$, amplitude $= 13$

(b) Period $= 360°$, amplitude $= 25$

(c) Period $= 360°$, amplitude $= 1.41 \left(\text{in fact } \sqrt{2}\right)$

B5 (a) $360°$

(b) $\sqrt{13} = 3.61$ (to 2 d.p.)

(c)

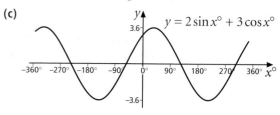

(d) Period $360°$, amplitude $\sqrt{a^2 + b^2}$

B6 (a)

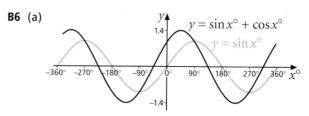

(b) $y = \sqrt{2} \sin(x + 45)°$

(c) $y = \sqrt{2} \sin(x + 45)°$
$= \sqrt{2}(\sin x° \cos 45° + \cos x° \sin 45°)$
$= \sqrt{2}\left(\sin x° \dfrac{1}{\sqrt{2}} + \cos x° \dfrac{1}{\sqrt{2}}\right)$
$= \sin x° + \cos x°$

B7 Taking $r = -\sqrt{13}$ we have
$\cos \alpha° = -\dfrac{2}{\sqrt{13}}$ and $\sin \alpha° = -\dfrac{3}{\sqrt{13}}$
$\implies \tan \alpha° = \frac{3}{2}$ (and α is in the third quadrant)
$\implies \alpha° = 236°$ (to the nearest degree)
Hence $2\sin x° + 3\cos x° = -\sqrt{13} \sin(x + 236)°$
This has a maximum when $\sin(x + 236)° = -1$,
i.e. when $(x + 236)° = 270°$, giving $x° = 34°$.

B8 (a) $r\sin x° \cos \alpha° - r\cos x° \sin \alpha°$

(b) $4\sin x° - 3\cos x° = r\sin(x - \alpha)°$
$\implies r\cos \alpha° = 4$ and $r\sin \alpha° = 3$
$\implies r^2 = 4^2 + 3^2 = 25 \ (r > 0)$, so $r = 5$

(c) $\tan \alpha° = \dfrac{r\sin \alpha°}{r\cos \alpha°} = \frac{3}{4} = 0.75$, so $\alpha° = 37°$

(d) $4\sin x° - 3\cos x° = 5\sin(x - 37)°$

(e) (i) -5

(ii) The minimum occurs when
$\sin(x - 37)° = -1$, i.e. $(x - 37)° = 270°$,
$x° = 307°$.

(f) A check with a graph plotter

Exercise B (p 93)

Angles in these answers are given to the nearest degree unless stated otherwise.

1 (a) $3\sin(\theta+30)°$ (b) $3\cos(\theta-60)°$

(c) $3\sin(\theta-330)°$ (d) $3\cos(\theta+300)°$

2 (a) $\sqrt{29}\sin(\theta+22)° \approx 5.4\sin(\theta+22)°$

(b) Maximum $= \sqrt{29} \approx 5.4$ when $\theta° = 68°$

3 (a) $\sqrt{5}\sin(\theta+63)° \approx 2.2\sin(\theta+63)°$

(b) $\sqrt{5}\cos(\theta-27)° \approx 2.2\cos(\theta-27)°$

(c) $\sqrt{5}\cos(\theta-27)° = \sqrt{5}\sin(90-(\theta-27))°$

$= \sqrt{5}\sin(117-\theta)° = -\sqrt{5}\sin(\theta-117)°$

$= \sqrt{5}\sin(\theta-117+180)° = \sqrt{5}\sin(\theta+63)°$

4 (a) $\sqrt{13}\sin(\theta-56)°$ (b) $-\sqrt{13}\cos(\theta+34)°$

(c) $\theta = 72°$

5 (a) $x° = 84°, 346°$ (b) $x° = 77°, 210°$

(c) $x° = 38°, 264°$

6 $\cos(x+45)° = 2\cos(x-45)°$

$\Rightarrow \cos x° \cos 45° - \sin x° \sin 45°$

$= 2(\cos x° \cos 45° + \sin x° \sin 45°)$

$\Rightarrow \cos x° \dfrac{1}{\sqrt{2}} - \sin x° \dfrac{1}{\sqrt{2}} = 2\left(\cos x° \dfrac{1}{\sqrt{2}} + \sin x° \dfrac{1}{\sqrt{2}}\right)$

$\Rightarrow \cos x° - \sin x° = 2(\cos x° + \sin x°)$

$\Rightarrow \cos x° - \sin x° = 2\cos x° + 2\sin x°$

$\Rightarrow -3\sin x° = \cos x°$

$\Rightarrow \dfrac{\sin x°}{\cos x°} = -\dfrac{1}{3}$, i.e. $\tan x° = -\dfrac{1}{3}$

7 (a) $2\sin\left(\theta - \dfrac{\pi}{6}\right)$ (b) $\theta = \dfrac{\pi}{3}$ or π

8 (a) $\tan(\theta-45°) = \dfrac{\tan\theta° - \tan 45°}{1 + \tan\theta° \tan 45°}$

$= \dfrac{\tan\theta° - 1}{1 + \tan\theta° \times 1} = \dfrac{\tan\theta° - 1}{1 + \tan\theta°}$

(b) $2 - \sqrt{3}$

9 (a) $\dfrac{12}{13}$ (b) $-\dfrac{16}{65}$

10 (a) (i) $13\cos(\theta+22.62)°$ (to 2 d.p.)

(ii) $\theta = 49.5°$ (to 1 d.p.)

(b) $\theta = 53.1°$ (to 1 d.p.)

Test yourself (p 94)

1 $\sin(x+90)° = \sin x° \cos 90° + \cos x° \sin 90° = \cos x°$

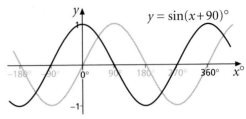

The graph of $y = \sin(x+90)°$ is a translation of $y = \sin x°$ (shown in grey above) by $\begin{bmatrix} -90° \\ 0 \end{bmatrix}$ and is clearly the same as $y = \cos x°$.

2 (a) $2\sin(\theta+345)°$ (b) $2\cos(\theta-105)°$

(c) $2\sin(\theta-15)°$ (d) $2\cos(\theta+255)°$

3 (a) $\cos 105° = \cos(60+45)°$

$= \cos 60° \cos 45° - \sin 60° \sin 45°$

$= \dfrac{1}{2} \times \dfrac{1}{\sqrt{2}} - \dfrac{\sqrt{3}}{2} \times \dfrac{1}{\sqrt{2}} = \dfrac{1-\sqrt{3}}{2\sqrt{2}}$

(b) $\sin 105° = \sin(60+45)°$

$= \sin 60° \cos 45° + \cos 60° \sin 45°$

$= \dfrac{\sqrt{3}}{2} \times \dfrac{1}{\sqrt{2}} + \dfrac{1}{2} \times \dfrac{1}{\sqrt{2}} = \dfrac{\sqrt{3}-1}{2\sqrt{2}}$

4 (a) $\dfrac{4}{5}$ (b) $\dfrac{7}{25}$ (c) $\dfrac{24}{25}$ (d) $\dfrac{24}{7}$

5 (a) $-\dfrac{5}{13}$ (b) $-\dfrac{119}{169}$ (c) $-\dfrac{169}{120}$ (d) $\dfrac{119}{120}$

6 $-\dfrac{1}{3} + \dfrac{\sqrt{10}}{3}$ or $-\dfrac{1}{3} - \dfrac{\sqrt{10}}{3}$

7 $\theta° = 27°, 90°, 207°$ or $270°$

8 (a) $x = \dfrac{\pi}{2}$ or $\dfrac{3\pi}{2}$ (b) $x = 0, \dfrac{3\pi}{4}, \pi, \dfrac{7\pi}{4}$ or 2π

(c) $x = 2.33$ or 3.96 (to 2 d.p.)

9 $\cot\dfrac{\theta}{2} - \tan\dfrac{\theta}{2} = \dfrac{1}{\tan\dfrac{\theta}{2}} - \tan\dfrac{\theta}{2} = \dfrac{1-\tan^2\dfrac{\theta}{2}}{\tan\dfrac{\theta}{2}}$

$= 2\left(\dfrac{1-\tan^2\dfrac{\theta}{2}}{2\tan\dfrac{\theta}{2}}\right) = 2\left(\dfrac{1}{\tan\left(2\times\dfrac{\theta}{2}\right)}\right) = \dfrac{2}{\tan\theta} = 2\cot\theta$

10 $\tan(A+B) + \tan(A-B)$

$$= \frac{\tan A + \tan B}{1 - \tan A \tan B} + \frac{\tan A - \tan B}{1 + \tan A \tan B}$$

$$= \frac{(\tan A + \tan B)(1 + \tan A \tan B) + (\tan A - \tan B)(1 - \tan A \tan B)}{(1 - \tan A \tan B)(1 + \tan A \tan B)}$$

$$= \frac{2\tan A + 2\tan A \tan^2 B}{1 - \tan^2 A \tan^2 B}$$

$$= 2\tan A \frac{1 + \tan^2 B}{\tan^2 A \left(\dfrac{1}{\tan^2 A} - \tan^2 B\right)}$$

$$= \frac{2\tan A}{\tan^2 A} \times \frac{\sec^2 B}{\cot^2 A - \tan^2 B}$$

$$= 2\cot A \frac{\sec^2 B}{\cot^2 A - \tan^2 B}$$

11 (a) $r\sin x° \cos\alpha° + r\cos x° \sin\alpha°$

(b) $\sqrt{34}\,\sin(x + 59)°$

(c) Maximum $= \sqrt{34}$ at $x = 31°$

12 (a) $5\sin(x + 53)°$

(b) Maximum $= \frac{2}{3}$ when $x° = 217°$

Minimum $= \frac{1}{4}$ when $x° = 37°$

13 (a) (i) $\sqrt{41}\,\sin(\theta - 39)°$

(ii) $\theta° = 67°$ or $191°$

(b) (i) $\sqrt{41}\,\cos(\theta + 51)°$

(ii) $\theta° = 67°$ or $191°$

(c) The answers are identical because the two equations being solved in (a) and (b) are equivalent.

14 (a) 1.00 or 5.77

(b) 2.14 or 3.65

(c) 1.19 or 4.45

(d) 1.57 or 5.76 $\left(\dfrac{\pi}{2} \text{ or } \dfrac{11\pi}{6}\right)$

(e) 1.57 $\left(\dfrac{\pi}{2}\right)$ or 5.94

(f) 0.50 or 5.10

7 Natural logarithms and e^x

A Introducing e (p 96)

A1 £2.00

A2 After six months the amount will be £1.00×1.5 = £1.50. After another six months it will be £1.50×1.5 = £2.25.

A3 After the first calculation the amount in pounds will be $1 \times 1\frac{1}{3}$. After the second calculation the amount in pounds will be $1 \times 1\frac{1}{3} \times 1\frac{1}{3} = \left(1\frac{1}{3}\right)^2$. After the third and final calculation the amount in pounds will be $\left(1\frac{1}{3}\right)^3 = \left(1 + \frac{1}{3}\right)^3$ as required. This amount is £2.37 to the nearest penny.

A4 (a) After the first calculation the amount in pounds will be $1 \times \left(1 + \dfrac{1}{n}\right)$. After the second calculation the amount in pounds will be $\left(1 + \dfrac{1}{n}\right)^2$ and so on. After the nth and final calculation the amount in pounds will be $\left(1 + \dfrac{1}{n}\right)^n$ as required.

(b) £2.70 to the nearest penny

(c) Comments such as:
The amount increases but at a slower and slower rate. The amount appears to converge to a limit of £2.72.

A5 (a) A sketch of $y = \left(1 + \dfrac{1}{x}\right)^x$ is

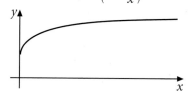

(b) 2.718

A6 (a) $f(3) = \dfrac{1}{0!} + \dfrac{1}{1!} + \dfrac{1}{2!} + \dfrac{1}{3!}$
$= \frac{1}{1} + \frac{1}{1} + \frac{1}{2} + \frac{1}{6} = 2\frac{2}{3}$

(b) (i) 2.716667   (ii) 2.718279

(iii) 2.718282   (iv) 2.718282

(c) They are all very close and f(10) and f(15) agree to six decimal places.

Exercise A (p 97)

1 (a) F　　(b) T　　(c) T　　(d) T　　(e) T
　(f) T　　(g) F　　(h) T　　(i) T　　(j) T
　(k) F　　(l) T

2 (a) 7.3891　　(b) 0.0498　　(c) 1.6487
　(d) 0.7788　　(e) 0.6796

3 C, A, D, B

B Natural logarithms (p 97)

B1 (a) 1.95　　(b) 0.69

B2 (a) 0.693 147
　(b) $e^{\ln 2} = e^{0.693\,147} = 1.999\,999\,639 \ldots \approx 2$

B3 4

B4 (a) 4.000　　(b) 1.4

B5 (a) $x = 2.3026$　　(b) $x = 0.9163$
　(c) $x = -0.6931$　　(d) $x = -2.3026$

B6 e^x is positive for all x so no value for x exists such that $e^x = -5$.

B7 (a) e^{-3}　　(b) -3

B8 (a) 5　　(b) 1　　(c) -2　　(d) 0.5　　(e) 0

B9 No value for x exists such that $e^x = -2$ so $\ln(-2)$ does not exist.

B10 $2\ln 3 + 5\ln 2 = \ln 3^2 + \ln 2^5 = \ln 9 + \ln 32$
　$= \ln(9 \times 32) = \ln 288$ as required

B11 (a) $\ln 6$　　(b) $\ln 2$　　(c) $\ln 36$　　(d) $\ln 128$

B12 (a) 1.609 438
　(b) (i) 3.2　　(ii) 2.6　　(iii) -0.6

B13 $\ln(8e^2) = \ln 8 + \ln e^2 = \ln 2^3 + 2 = 2 + 3\ln 2$ as required

Exercise B (p 100)

1 (a) $x = 1.099$　　(b) $x = 1.386$　　(c) $x = 1.253$
　(d) $x = 1.996$　　(e) $x = -0.536$　　(f) $x = 4.159$
　(g) $x = 0.101$　　(h) $x = 0.564$　　(i) $x = -4$

2 (a) (i) 11　　(ii) 132　　(iii) 88106
　(b) Just over 4.8 hours

3 (a) 5 mg
　(b) (i) 4.09 mg　　(ii) 1.84 mg　　(iii) 0.04 mg
　(c) Just over 31 hours

4 One method is to solve the equation $e^{3x-2} = 5$ directly as follows.
$$e^{3x-2} = 5 \implies 3x - 2 = \ln 5$$
$$\implies 3x = \ln 5 + 2$$
$$\implies x = \tfrac{1}{3}(\ln 5 + 2) \text{ as required}$$

5 (a) $x = \tfrac{1}{2}(\ln 5 - 1)$　(b) $x = \tfrac{4}{3}$　(c) $x = 2(1 - \ln 4)$

6 (a) $\ln \tfrac{1}{2} = \ln 2^{-1} = -\ln 2$
　(b) $e^{-\ln 4} = e^{\ln 4^{-1}} = e^{\ln \frac{1}{4}} = \tfrac{1}{4}$ or $e^{-\ln 4} = \dfrac{1}{e^{\ln 4}} = \tfrac{1}{4}$

7 $x = \tfrac{1}{2}\ln 3$

8 $x = \tfrac{1}{5}\ln 6 \left(\text{or } -\tfrac{1}{5}\ln\tfrac{1}{6}\right)$

9 (a) $x = 54.598$　　(b) $x = 0.025$　　(c) $x = 5.294$
　(d) $x = 0.822$　　(e) $x = 3.741$　　(f) $x = 2.096$

10 One method is to solve the equation $\ln(4x - 1) = 2$ directly as follows.
$$\ln(4x - 1) = 2 \implies 4x - 1 = e^2$$
$$\implies 4x = e^2 + 1$$
$$\implies x = \tfrac{1}{4}(e^2 + 1) \text{ as required}$$

11 (a) $x = \tfrac{1}{4}e^5$　　(b) $x = e^{-2} + 1$　　(c) $x = \tfrac{1}{2}(e - 3)$

12 (a) $\ln(x^3)$　　(b) $\ln(20x)$　　(c) $\ln(x^2)$
　(d) $\ln(25x)$　　(e) $\ln\left(\tfrac{1}{2}x\right)$　　(f) $\ln\dfrac{3}{x}$

13 (a) $x = \tfrac{1}{5}e^2$　　(b) $x = 3e$　　(c) $x = \sqrt{\tfrac{1}{2}e^3}$
　(d) $x = \tfrac{1}{40}$　　(e) $x = \tfrac{1}{2}e^5 - 1$　　(f) $x = 5 - 4e$

14 (a)　$\ln(x + 2) + \ln x = \ln 8$
　$\implies$　$\ln(x(x + 2)) = \ln 8$
　$\implies$　$x^2 + 2x = 8$
　$\implies$　$x^2 + 2x - 8 = 0$
　so x must be a solution of this equation.
　(b) The two solutions are $x = 2$ and $x = -4$. Only $x = 2$ is a solution of $\ln(x + 2) + \ln x = \ln 8$ as $\ln x$ is not defined for a negative number.

15 (a) $x = \tfrac{3}{4}$　　(b) $x = 2$　　(c) $x = 1, 5$　　(d) $x = \tfrac{3}{5}$

16 (a)　　　　$e^{2x} - 4e^x + 3 = 0$
　$\implies$　$(e^x)^2 - 4(e^x) + 3 = 0$
　$\implies$　$y^2 - 4y + 3 = 0$ where $y = e^x$
　(b) $x = 0, \ln 3$

17 (a) $x = 0, \ln 5$ (b) $x = \ln 2$ (c) $x = \ln 4$

18 $x = 2 + \sqrt{3}, 2 - \sqrt{3}$

19 (a) $x = \ln\left(\frac{2}{3}\right)$ (b) $x = \ln 3 - 1$ (c) $x = \ln 4, 0$

C Graphs (p 102)

C1 $y > 0$

C2 $x > 0$

C3 (a)

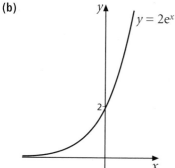
$y = e^{-x}$

(b)

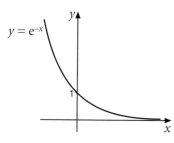
$y = 2e^x$

(c)

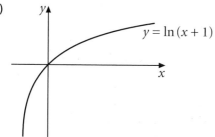
$y = \ln(x + 1)$

(d)

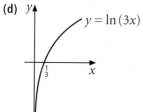
$y = \ln(3x)$

C4 (a) $f(x) > 0$ (b) $f(x) > 0$

(c) $f(x) \in \mathbb{R}$ (d) $f(x) \in \mathbb{R}$

Exercise C (p 104)

1 (a) 6

(b)

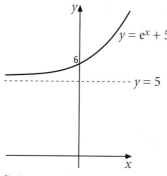
$y = e^x + 5$
$y = 5$

(c) $f(x) > 5$

2

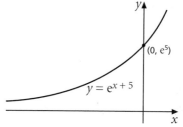
$(0, e^5)$
$y = e^{x+5}$

3 A stretch in the x-direction of factor 3

4 (a) A translation of $\begin{bmatrix} 0 \\ -1 \end{bmatrix}$

(b) $y = 0 \Rightarrow \ln x - 1 = 0 \Rightarrow \ln x = 1 \Rightarrow x = e^1 = e$. So the graph meets the x-axis at $(e, 0)$.

5 (a)

$(e^3, 0)$
$y = \ln x - 3$

(b)

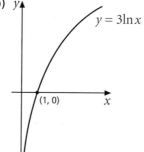
$y = 3\ln x$
$(1, 0)$

(c)

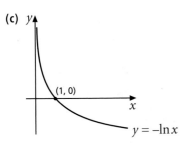

$y = -\ln x$

(d)

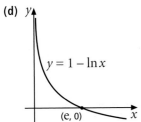

$y = 1 - \ln x$

6 A reflection in the y-axis and a stretch of factor 5 in the y-direction (or vice versa)

7

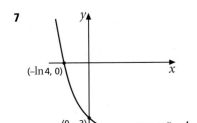

$y = e^{-x} - 4$

8 (a) -4

(b)

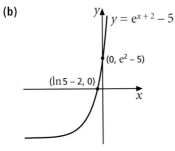

$y = e^{x+2} - 5$

(c) $y = -5$

(d) $f(x) > -5$

(e) $x = \ln 6 - 2$

9 (a) A translation of $\begin{bmatrix} -2 \\ 0 \end{bmatrix}$

(b) $x = -2$

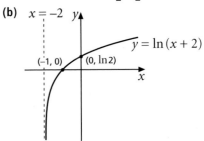

$y = \ln(x + 2)$

10 (a) 1

(b) $f(x) = 0 \Rightarrow 2e^{-x} - 1 = 0$
$\Rightarrow e^{-x} = \frac{1}{2}$
$\Rightarrow -x = \ln\frac{1}{2} = \ln 2^{-1} = -\ln 2$
$\Rightarrow x = \ln 2$

(c) $f(x) > -1$

(d) $x = -\ln 2 \left(\text{or } \ln\frac{1}{2} \right)$

11 (a) $p = 3, q = \frac{1}{2}(e^{-3} - 1)$ **(b)** $x = -\frac{1}{2}$

(c) $x = \frac{1}{2}(e^{-5} - 1)$

12

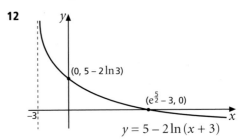

$y = 5 - 2\ln(x + 3)$

13 (a)

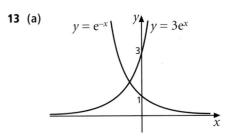

(b) (i) $3e^x = e^{-x}$
$\Rightarrow 3e^x \times e^x = e^{-x} \times e^x$
$\Rightarrow 3e^{2x} = 1$
$\Rightarrow e^{2x} = \frac{1}{3}$

so x must be a root of this equation.

(ii) $x = -0.5493$ (to 4 s.f.)

14 (a)

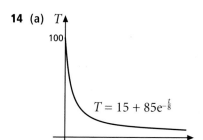

$T = 15 + 85e^{-\frac{t}{8}}$

(b) About 10 minutes

15 (a) 2

(b) (i) 4.39 **(ii)** 0.28 **(iii)** 2.95

(c) $x = \ln 3$

(d)

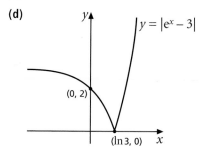

$y = |e^x - 3|$

(0, 2)

(ln 3, 0)

(e) $f(x) \geq 0$

(f) The line $y = 1$ cuts the graph above in two places, so the equation $f(x) = 1$ has two solutions; $x = \ln 4, \ln 2$

(g) $x = \ln 8$

16 (a) $g(3) = g(1) = 0$

(b)

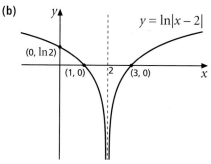

$y = \ln|x - 2|$

(0, ln2)

(1, 0) 2 (3, 0)

(c) $x = 2 + e^3, 2 - e^3$

D Inverses

Exercise D (p 108)

1 (a) $\frac{1}{2}\ln x$ **(b)** $\ln x - 2$ **(c)** $\ln(x - 2)$
(d) $\ln\left(\frac{1}{2}x\right)$ **(e)** $2 - \ln x$ **(f)** $-\ln(x + 2)$

2 (a) $\frac{1}{5}e^x$ **(b)** $e^x - 5$ **(c)** e^{x-5}
(d) $e^{\frac{1}{5}x}$ **(e)** $5e^x$ **(f)** $e^{\frac{1}{5}(x+1)}$

3 (a) $f(x) > 2$ **(b)** $\ln(x - 2) + 1$
(c) Domain: $x > 2$; range: $f^{-1}(x) \in \mathbb{R}$
(d)

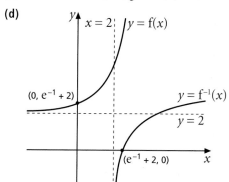

$x = 2$ $y = f(x)$

(0, $e^{-1} + 2$)

$y = f^{-1}(x)$

$y = 2$

($e^{-1} + 2$, 0)

4 (a) (i) $e^{x+2} - 1$
(ii) The solution is $g^{-1}(7)$, which is $e^9 - 1$.
(b) Domain: $x \in \mathbb{R}$; range: $g^{-1}(x) > -1$

5 (a) $\ln\left(\frac{1}{2}(x - 3)\right)$
(b) (i) $x > 3$
(ii) 1 is not in the domain of f^{-1} so $f(x) = 1$ has no solution.

6 (a) $\frac{1}{5}e^x + 2$
(b)

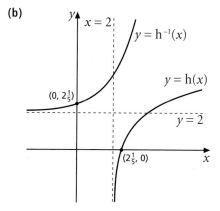

$x = 2$ $y = h^{-1}(x)$

(0, $2\frac{1}{5}$)

$y = h(x)$

$y = 2$

($2\frac{1}{5}$, 0)

7 (a) $3e^2$ **(b)** $-\ln\left(\frac{1}{3}x\right)$
(c) Domain: $0 < x \leq 3e^2$; range: $f^{-1}(x) \geq -2$

8 (a) $t = -\frac{10}{3}\ln\left(\frac{V}{14\,000}\right)$ **(b)** 2008

Mixed questions (p 109)

1 (a) (i) $(0, 40)$ **(ii)** $f(x) > 0$ **(iii)** $40 \times 10^{\frac{1}{50}}$

(b) (i) $g(x) > 1000$

(ii)

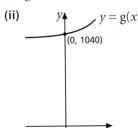

$y = g(x)$

$(0, 1040)$

(iii) $g(x) = 1200$

$$\Rightarrow \quad 40\left(25 + e^{\frac{x}{50}}\right) = 1200$$

$$\Rightarrow \quad 25 + e^{\frac{x}{50}} = 30$$

$$\Rightarrow \quad e^{\frac{x}{50}} = 5$$

$$\Rightarrow \quad \frac{x}{50} = \ln 5$$

$$\Rightarrow \quad x = 50\ln 5$$

(c) (i) 1040 **(ii)** 80.5 days

2 (a)

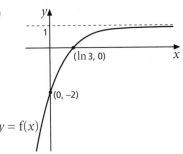

$(\ln 3, 0)$

$(0, -2)$

$y = f(x)$

(b) $x > \ln 3$

3 (a) $2(e^x + 3)$ **(b)** 4

(c) $gf(x) = 4e^{2\ln\left(\frac{1}{2}x - 3\right)}$

$$= 4e^{\ln\left(\frac{1}{2}x - 3\right)^2}$$

$$= 4\left(\frac{1}{2}x - 3\right)^2$$

$$= \left(2\left(\frac{1}{2}x - 3\right)\right)^2$$

$$= (x - 6)^2 \text{ as required}$$

4 (a) 184.73

(b) A translation of $\begin{bmatrix} 0 \\ 1 \end{bmatrix}$

(c) e^{x-1}

(d) $fg(x) = 1 + \ln(ex)^2$

$$= 1 + 2\ln(ex)$$

$$= 1 + 2(\ln e + \ln x)$$

$$= 1 + 2(1 + \ln x)$$

$$= 3 + 2\ln x \text{ as required}$$

(e)

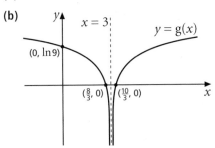

$y = |fg(x)|$

$(e^{-\frac{3}{2}}, 0)$

5 (a) 52.5

(b)

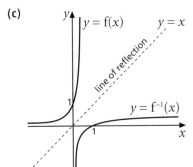

$x = 3$

$y = g(x)$

$(0, \ln 9)$

$\left(\frac{8}{3}, 0\right)$ $\left(\frac{10}{3}, 0\right)$

Test yourself (p 110)

1 (a) $\ln p + \ln q$ **(b)** $3\ln p + 2\ln q$

(c) $2\ln p - \ln q$ **(d)** $\ln p - \frac{1}{2}\ln q$

2 (a) 1 **(b)** $\frac{1}{4}\ln x$

(c)

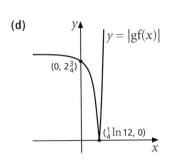

$y = f(x)$ $y = x$

line of reflection

$y = f^{-1}(x)$

(d)

$y = |gf(x)|$

$\left(0, 2\frac{3}{4}\right)$

$\left(\frac{1}{4}\ln 12, 0\right)$

3 (a) A stretch of factor $\frac{1}{4}$ in the x-direction

(b) A translation of $\begin{bmatrix} -1 \\ 0 \end{bmatrix}$

(c) A reflection in the x-axis

4 $y = \frac{1}{3}(5e^2 - 2)$

5 (a)

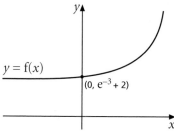

$y = f(x)$

$(0,\ e^{-3} + 2)$

(b) $f(x) > 2$

(c) $x = \ln 5 + 3$

(d) $\ln(x - 2) + 3$ with domain $x > 2$

6 (a) A stretch of factor 3 in the y-direction

(b) $e^{\frac{1}{3}x}$

7 $a = -3\frac{1}{2},\ b = -3,\ c = \ln 7$

8 (a) $2 - \frac{1}{2}e^x$

(b)

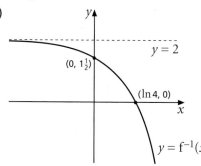

$y = 2$

$(0,\ 1\frac{1}{2})$

$(\ln 4,\ 0)$

$y = f^{-1}(x)$

(c) 0.369   **(d)** 2.141

9 (a) Reflection in the y-axis followed by a stretch of factor 3 in the y-direction (or vice versa)

(b) $x = 3.40$

(c) $-\ln\left(\frac{1}{3}x\right)$ or $\ln\left(\frac{3}{x}\right)$

(d) Domain: $x > 0$; range: $f^{-1}(x) \in \mathbb{R}$

(e)

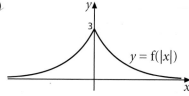

$y = f(|x|)$

8 Differentiation

A Exponential functions (p 112)

A1 (a) $\dfrac{2^h - 1}{h - 0} = \dfrac{2^h - 1}{h}$

(b) 0.7177

(c) 0.7053, 0.6956, 0.6934, 0.6932

(d) 0.69

A2 1.161, 1.105, 1.099, 1.099
The gradient is approximately 1.1.

A3 The results are 1.0517..., 1.0050..., 1.0005...

A4 (a) $\dfrac{e^{p+h} - e^p}{h} = \dfrac{e^p e^h - e^p}{h} = \dfrac{e^p(e^h - 1)}{h}$

(b) We already know that $\dfrac{e^h - 1}{h}$ approaches 1, and so the value of this expression gets closer and closer to $e^p \times 1 = e^p$.
The gradient of $y = e^x$ at $x = p$ is e^p.

A5 (a) $3e^x$ **(b)** $5e^x + 4$

(c) $2e^x - 2x$ **(d)** $e^x - \dfrac{2}{x^2}$

A6 $e^2 e^x = e^{x+2}$

Exercise A (p 114)

1 (a) $e^x - 5$ **(b)** $2x + 4e^x$

(c) $e^x - \dfrac{3}{2\sqrt{x}}$ **(d)** $3e^x + 3x^2$

2 (a) $2e^x + 2x^{-3}$ **(b)** $6e^x - \dfrac{1}{x^2}$

(c) $\frac{3}{2}x^{\frac{1}{2}} + 2e^x$ **(d)** $-\frac{1}{2}x^{-\frac{3}{2}} + 2e^x$

3 (a) $\dfrac{dy}{dx} = e^x - 3$

(b) When $x = \ln 3$, $\dfrac{dy}{dx} = e^{\ln 3} - 3 = 3 - 3 = 0$

(c) $y = e^{\ln 3} - 3\ln 3$
$= 3 - 3\ln 3$
$= -0.296$

(d) $\dfrac{d^2 y}{dx^2} = e^x$

(e) 3

(f) Minimum, as $\dfrac{d^2 y}{dx^2}$ is positive

4 $f'(x) = e^x + \dfrac{3}{x^4}$; e^x and $\dfrac{3}{x^4}$ are always positive, so the gradient of $f(x)$ is always positive.

B The derivative of ln x (p 115)

B1 (a) $\dfrac{1}{x} + 2x$ (b) $\dfrac{1}{x} - \dfrac{1}{2\sqrt{x}}$

(c) $\dfrac{1}{x} + 5e^x$ (d) $-\dfrac{1}{x^2} - \dfrac{1}{x}$

B2 $\ln ab = \ln a + \ln b$ so $\ln 3x = \ln 3 + \ln x$

The derivative of $\ln 3 + \ln x$ is

$0 + \dfrac{1}{x} = \dfrac{1}{x}$ ($\ln 3$ is a constant).

B3 (a) $\ln 5x^2 = \ln 5 + \ln x + \ln x = \ln 5 + 2\ln x$;

$\dfrac{d}{dx}(\ln 5 + 2\ln x) = 0 + \dfrac{2}{x} = \dfrac{2}{x}$

(b) (i) $\dfrac{3}{x}$ (ii) $\dfrac{1}{2x}$ (iii) $-\dfrac{3}{x}$ (iv) $-\dfrac{1}{x}$

Exercise B (p 116)

1 (a) $2x + \dfrac{1}{x}$ (b) $\dfrac{3}{x} - 3x^2$

(c) $\dfrac{1}{2\sqrt{x}} - \dfrac{2}{x}$ (d) $\dfrac{1}{x} + e^x$

2 $\ln 7 + 3\ln x$; $\dfrac{3}{x}$

3 (a) $\dfrac{2}{x}$ (b) $\dfrac{5}{x}$ (c) $\dfrac{1}{2x}$ (d) $\dfrac{3}{2x}$

4 $\ln 3 - 2\ln x$; $-\dfrac{2}{x}$

5 (a) $e^x + \dfrac{3}{x}$ (b) $-\dfrac{1}{x}$

(c) $\dfrac{2}{3x}$ (d) $-\tfrac{1}{2}x^{-\frac{3}{2}} - \tfrac{1}{2}x^{-1}$

6 (a) $\dfrac{dy}{dx} = \dfrac{1}{x} - 1$; at $x = 1$, $\dfrac{dy}{dx} = 0$

(b) $\dfrac{d^2y}{dx^2} = -\dfrac{1}{x^2} \le 0$

7 $\left(\tfrac{1}{2}, \tfrac{1}{16} - \tfrac{1}{4}\ln\tfrac{1}{2}\right)$; minimum

C Differentiating a product of functions (p 117)

C1 (a) $6x^2 + 2x$

(b) $2x \times 2 = 4x$, not $6x^2 + 2x$

C2 $y + \delta y = (u + \delta u)(v + \delta v)$

$\qquad = uv + u\delta v + v\delta u + \delta u\delta v$

$\qquad = y + u\delta v + v\delta u + \delta u\delta v$

$\quad \delta y = u\delta v + v\delta u + \delta u\delta v$

C3 $\dfrac{du}{dx} = 2x$, $\dfrac{dv}{dx} = 2$; $6x^2 + 2x$

C4 (a) $f'(x) = 3x^2$, $g'(x) = 2x$; $5x^4 - 9x^2$

(b) $f(x)g(x) = x^5 - 3x^3$; $5x^4 - 9x^2$

C5 $3x^2e^x + x^3e^x$

C6 $4x^3\ln x + x^3 = x^3(4\ln x + 1)$

C7 (a) $(x^2 - x - 3)e^x$ (b) $(2x - 3)\ln x + x - 3$

(c) $x^{-\frac{1}{2}}\left(\tfrac{1}{2}\ln x + 1\right)$

Exercise C (p 119)

1 (a) $(5 + x)e^xx^4$ (b) $\left(\tfrac{1}{2}x^{-\frac{1}{2}} + x^{\frac{1}{2}}\right)e^x$

(c) $x(2\ln x + 1)$ (d) $3e^x(x + 1)$

(e) $e^x\left(\ln x + \dfrac{1}{x}\right)$ (f) $e^x\sqrt[3]{x}\left(\dfrac{1}{3x} + 1\right)$

(g) $\dfrac{1 - 2\ln x}{x^3}$ (h) $e^xx^{-3}(x - 2)$

2 (a) $e^x(2x + 5)$

(b) $4\ln x + 4 - \dfrac{1}{x}$

(c) $e^x(x + 1)^2$

(d) $\dfrac{1}{\sqrt{x}}\left(\tfrac{1}{2}\ln x + 1\right) - \dfrac{1}{x}$

(e) $e^x\left(1 - \dfrac{1}{x} + \dfrac{1}{x^2}\right)$

(f) $e^x\left(x + 1 + \dfrac{1}{x} + \ln x\right)$

3 (a) $e^x(x^3 + 4x^2 + 2x)$

(b) $(3x^2 + x^3)e^x$; $(2x + x^2)e^x$; $e^x(x^3 + 4x^2 + 2x)$

(c) $(3x^2 + 2x)\ln x + x^2 + x$

4 (a) $\dfrac{dy}{dx} = e^x(1 + x)$

(b) At $x = -1$, $\dfrac{dy}{dx} = 0$

(c) $y = -\dfrac{1}{e}$

(d) $\dfrac{d^2y}{dx^2} = e^x(2 + x)$; minimum

5 (a) $e^x(x^2 - 5x + 6)$

(b) $e^x(x - 2)(x - 3)$

(c) $x = 2, x = 3$

(d) $f''(x) = e^x(x^2 - 3x + 1)$
At $x = 2$, $f(x)$ is a maximum.
At $x = 3$, $f(x)$ is a minimum.

6 (a) $\dfrac{dy}{dx} = \ln x + 1$; at $x = e$, $\dfrac{dy}{dx} = 2$,

so the gradient is 2

(b) At $x = \dfrac{1}{e}$, $\dfrac{dy}{dx} = -1 + 1 = 0$

(c) $\dfrac{d^2y}{dx^2} = \dfrac{1}{x}$; positive at $x = \dfrac{1}{e}$

7 (a) $\dfrac{dy}{dx} = x^2 \times \dfrac{1}{x} + \ln x \times 2x$

$= x + 2x \ln x$

$= x(1 + 2\ln x)$

(b) $x(1 + 2\ln x) = 0$ at a stationary point

$\Rightarrow x = 0$ or $1 + 2\ln x = 0$

y is not defined when $x = 0$

so use $1 + 2\ln x = 0$

$\Rightarrow \qquad 2\ln x = -1$

$\Rightarrow \qquad \ln x = -0.5$

$\Rightarrow \qquad x = e^{-0.5}$

(c) Minimum

D Differentiating a quotient (p 120)

D1 (a) $\dfrac{x^2 + 2x}{(x+1)^2}$ **(b)** $\dfrac{-2x^3 - 1}{(x^3 - 1)^2}$

(c) $\dfrac{1 - \ln x}{x^2}$ **(d)** $\dfrac{e^x(x - 2)}{(x - 1)^2}$

Exercise D (p 121)

1 (a) $\dfrac{2x^2 + 6x}{(2x + 3)^2}$ **(b)** $\dfrac{2x^3 - 3x^2}{(x - 1)^2}$

(c) $\dfrac{1 - x^2}{(x^2 + 1)^2}$ **(d)** $\dfrac{2 - 2x^2}{(x^2 + x + 1)^2}$

(e) $\dfrac{2x - x^2}{(x^2 - x + 1)^2}$ **(f)** $\dfrac{-4x^3 - 1}{(2x^3 - 1)^2}$

(g) $\dfrac{2x - x^4}{(x^3 + 1)^2}$ **(h)** $\dfrac{-x^4 + 8x^2 + 2x + 5}{(x^3 + 5x + 1)^2}$

2 (a), (b) $e^x\left(\dfrac{1}{x^2} - \dfrac{2}{x^3}\right)$

3 (a) $\dfrac{xe^x}{(x + 1)^2}$ **(b)** $\dfrac{e^x(x - 1)^2}{(x^2 + 1)^2}$

(c) $\dfrac{2x + 2xe^x - x^2e^x}{(e^x + 1)^2}$ **(d)** $\dfrac{x - 1 - x\ln x}{x(x - 1)^2}$

(e) $\dfrac{x^2(1 - 2\ln x) + 1}{x(x^2 + 1)^2}$ **(f)** $\dfrac{e^x(x\ln x - 1)}{x(\ln x)^2}$

4 (a) $\dfrac{dy}{dx} = \dfrac{(x + 1) \times 1 - x \times 1}{(x + 1)^2}$

$= \dfrac{1}{(x + 1)^2}$

(b) $\dfrac{1}{(x + 1)^2} = 0$ has no solution.

5 (a) $f'(x) = \dfrac{(x^2 + 3) \times 1 - (x + 1) \times 2x}{(x^2 + 3)^2}$

$= \dfrac{x^2 + 3 - 2x^2 - 2x}{(x^2 + 3)^2}$

$= \dfrac{-(x^2 + 2x - 3)}{(x^2 + 3)^2}$

$= \dfrac{-(x - 1)(x + 3)}{(x^2 + 3)^2}$

(b) $x = 1$, $x = -3$

6 $x = 2$, $x = -2$

7 $(3, e^3)$

8 (a) (i) $\dfrac{1 - \ln x}{x^2}$ **(ii)** $\dfrac{2\ln x - 3}{x^3}$

(b) $\left(e, \dfrac{1}{e}\right)$

(c) Maximum

9 $\dfrac{dy}{dx} = \dfrac{\ln x \times 2x - x^2 \times \dfrac{1}{x}}{(\ln x)^2}$

$= \dfrac{2x\ln x - x}{(\ln x)^2}$

At a stationary point $\dfrac{dy}{dx} = 0$

so the numerator must be 0.

$2x\ln x - x = 0$

$x(2\ln x - 1) = 0$

y is not defined for $x = 0$

so use $\quad 2\ln x - 1 = 0$

$\Rightarrow \qquad 2\ln x = 1$

$\Rightarrow \qquad \ln x = 0.5$

$\Rightarrow \qquad x = e^{0.5}$

When $x = e^{0.5}$, $y = \dfrac{\left(e^{0.5}\right)^2}{\ln e^{0.5}}$

$\qquad = \dfrac{e}{0.5}$

$\qquad = 2e$

E Differentiating sin x, cos x and tan x (p 122)

E1 0.9983, 1.0000, 1.0000; the gradient appears to be 1.

E2 (a) 0, −1, 0, 1

(b)
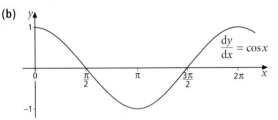

$\dfrac{dy}{dx} = \cos x$

E3
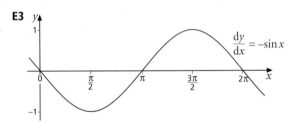

$\dfrac{dy}{dx} = -\sin x$

E4 (a) $\cos x + 3 + 2x$ (b) $\tfrac{1}{2}x^{-\frac{1}{2}} + \sin x$

(c) $3e^x + 4\cos x$ (d) $\dfrac{5}{x} + 2\sin x$

E5 (a) $\dfrac{dy}{dx} = \dfrac{\cos x \times \cos x - \sin x \times (-\sin x)}{(\cos x)^2}$

$\qquad = \dfrac{\cos^2 x + \sin^2 x}{\cos^2 x}$

(b) Since $\cos^2 x + \sin^2 x = 1$

$\qquad \dfrac{dy}{dx} = \dfrac{1}{\cos^2 x} = \left(\dfrac{1}{\cos x}\right)^2 = \sec^2 x$

E6 It tends to infinity; the graph of $y = \tan x$ has a vertical asymptote.

Exercise E (p 123)

1 (a) $\cos x - \sin x$ (b) $\cos x - 2x$

(c) $3\cos x + 2\sin x$ (d) $1 + 2\sec^2 x$

2 (a) $3\cos x - 3x\sin x$ (b) $-\dfrac{1}{x^2}(x\sin x + \cos x)$

(c) $\dfrac{1}{x^3}(x\cos x - 2\sin x)$ (d) $e^x(\sec^2 x + \tan x)$

3 $f'(x) = \dfrac{x \times \cos x - \sin x \times 1}{x^2}$

$\qquad = \dfrac{\cos x}{x} - \dfrac{\sin x}{x^2}$

4 (a) $f'(x) = 3\cos x - 4\sin x$

(b) $0 = 3\cos x - 4\sin x$

$\qquad \Rightarrow \tan x = \tfrac{3}{4}$

$\qquad x = 0.644, \, 3.785$

5 $f'(x) = \dfrac{x^2 \times (-\sin x) - \cos x \times 2x}{\left(x^2\right)^2}$

$\qquad = -\dfrac{\sin x}{x^2} - \dfrac{2\cos x}{x^3}$

$\qquad f'\left(\dfrac{\pi}{2}\right) = -\dfrac{1}{\left(\dfrac{\pi}{2}\right)^2} - 0$

$\qquad = -\dfrac{4}{\pi^2}$

6 $0, \, 2\pi$

7 $\dfrac{dy}{dx} = \dfrac{\sin x \times (-\sin x) - \cos x \times \cos x}{\sin^2 x}$

$\qquad = \dfrac{-\left(\sin^2 x + \cos^2 x\right)}{\sin^2 x}$

$\qquad = -\dfrac{1}{\sin^2 x}$

$\qquad = -\text{cosec}^2 x$

F Differentiating a function of a function (p 124)

F1 (a) $\cos u$ (b) $2x$

(c) $\dfrac{dy}{dx} = \dfrac{dy}{du} \times \dfrac{du}{dx} = \cos u \times 2x = 2x\cos\left(x^2 - 3\right)$

F2 $2\sin x \cos x$

F3 $\dfrac{dy}{dx} = \dfrac{dy}{du} \times \dfrac{du}{dx} = \tfrac{1}{2}u^{-\frac{1}{2}} \times 2x$

$\qquad = x\left(x^2 + 1\right)^{-\frac{1}{2}}$

$\qquad = \dfrac{x}{\sqrt{x^2 + 1}}$

F4 (a) $-2\sin 2x$ (b) $-3e^{-3x}$

(c) $\dfrac{6}{6x + 1}$ (d) $\dfrac{-1}{\sqrt{5 - 2x}}$

(e) $\dfrac{-4}{(3 + 4x)^2}$ (f) $5\cos(5x - 2)$

(g) $2(3x-2)^{-\frac{1}{3}}$ **(h)** $-2(1+4x)^{-\frac{3}{2}}$

F5 $\dfrac{x\cos x^2}{\sqrt{\sin x^2}}$

Exercise F (p 125)

1 (a) $3\cos(3x-2)$ **(b)** $-5\sin(5x+1)$

(c) $-2\cos(3-2x)$ **(d)** $-\frac{1}{2}\sin\frac{1}{2}x$

(e) $\dfrac{3}{3x+1}$ **(f)** $\dfrac{5}{2\sqrt{5x-1}}$

(g) $14(2x+1)^6$ **(h)** $4\sec^2 4x$

(i) $\frac{2}{3}(2x+1)^{-\frac{2}{3}}$ **(j)** $3\sec^2(3x+2)$

(k) $5e^{5x-2}$ **(l)** $\frac{1}{2}\sin(1-\frac{1}{2}x)$

2 $\dfrac{dy}{du}=2u,\ \dfrac{du}{dx}=-\sin x,\ \dfrac{dy}{dx}=-2\sin x\cos x$

3 $\dfrac{dy}{du}=\dfrac{1}{u},\ \dfrac{du}{dx}=\cos x,\ \dfrac{dy}{dx}=\cot x$

4 $e^{\sin x}\cos x$

5 (a) $e^x\cos(e^x)$ **(b)** $\dfrac{2\ln x}{x}$

(c) $2\tan x\sec^2 x$ **(d)** $10x(x^2+1)^4$

6 (a) $e^{\tan x}\sec^2 x$ **(b)** $\dfrac{1}{2x\sqrt{\ln x}}$

(c) $\dfrac{-1}{x(\ln x)^2}$ **(d)** $\dfrac{\cos(\sqrt{x})}{2\sqrt{x}}$

7 (a) $\dfrac{2x}{x^2+1}$ **(b)** $\dfrac{2\ln x}{x}$

(c) $2xe^{x^2}$ **(d)** $e^x\sec^2 e^x$

(e) $-2x\cos(1-x^2)$ **(f)** $\dfrac{x}{\sqrt{1+x^2}}$

(g) $\dfrac{-2x}{(1+x^2)^2}$ **(h)** $\dfrac{-6x}{(1+x^2)^4}$

8 (a) Let $u=1-x^2$ so $y=u^{-\frac{1}{2}}$

$$\dfrac{dy}{dx}=\dfrac{dy}{du}\times\dfrac{du}{dx}$$

$$=-\tfrac{1}{2}u^{-\frac{3}{2}}(-2x)$$

$$=\dfrac{x}{(1-x^2)^{\frac{3}{2}}}$$

(b) 1.172 (to 4 s.f.)

G Further trigonometrical functions (p 126)

G1 (a) $\dfrac{dy}{du}=-u^{-2},\ \dfrac{du}{dx}=\cos x$

$\dfrac{dy}{dx}=\dfrac{dy}{du}\times\dfrac{du}{dx}=-\dfrac{1}{u^2}\cos x=-\dfrac{1}{\sin^2 x}\cos x$

(b) $\dfrac{-\cos x}{\sin^2 x}=-\dfrac{1}{\sin x}\times\dfrac{\cos x}{\sin x}=-\text{cosec}\,x\cot x$

G2 $\dfrac{dy}{du}=-u^{-2},\ \dfrac{du}{dx}=-\sin x$

$\dfrac{dy}{dx}=\dfrac{dy}{du}\times\dfrac{du}{dx}=\dfrac{1}{u^2}\sin x=\dfrac{\sin x}{\cos^2 x}=\tan x\sec x$

G3 $\dfrac{dy}{du}=-u^{-2},\ \dfrac{du}{dx}=\sec^2 x$

$\dfrac{dy}{dx}=-\dfrac{1}{\tan^2 x}\sec^2 x=-\dfrac{\cos^2 x\sec^2 x}{\sin^2 x}$

$=-\dfrac{1}{\sin^2 x}=-\text{cosec}^2 x$

G4 $\dfrac{d}{dx}(\cot x)=\dfrac{-\sin^2 x-\cos^2 x}{\sin^2 x}=\dfrac{-1}{\sin^2 x}=-\text{cosec}^2 x$

G5 (a) $\dfrac{dx}{dy}=\cos y$

(b) $\dfrac{dy}{dx}=\dfrac{1}{\cos y}=\dfrac{1}{\sqrt{1-\sin^2 y}}=\dfrac{1}{\sqrt{1-x^2}}$

Exercise G (p 128)

1 (a) $-2\,\text{cosec}\,2x\cot 2x$

(b) $-4\,\text{cosec}^2 4x$

(c) $3\sec(3x-1)\tan(3x-1)$

(d) $-2x\,\text{cosec}(x^2+1)\cot(x^2+1)$

(e) $\dfrac{\text{cosec}^2\left(\frac{1}{x}\right)}{x^2}$

(f) $-\dfrac{\text{cosec}(\sqrt{x})\cot(\sqrt{x})}{2\sqrt{x}}$

(g) $-3x^2\sec(1-x^3)\tan(1-x^3)$

(h) $-(2x+1)\,\text{cosec}^2(x^2+x)$

2 $\dfrac{dy}{dx}=\sec x\tan x-\text{cosec}\,x\cot x$

When $x=\dfrac{\pi}{4},\ \dfrac{dy}{dx}=\sqrt{2}-\sqrt{2}=0$

3 $\dfrac{dy}{dx}=1+\text{cosec}^2 x\geq 1$ for all values of x.

4 $\dfrac{dy}{dx}=2\sec x\times\sec x\tan x=2\sec^2 x\tan x$

5 (a) $-2\csc^2 x \cot x$ **(b)** $-2\csc^2 x \cot x$

  (c) $3\sec^3 x \tan x$ **(d)** $-6\csc^2 3x \cot 3x$

  (e) $-\dfrac{\csc^2 x}{2\sqrt{\cot x}}$ **(f)** $e^{\sec x}\sec x \tan x$

  (g) $e^x \sec(e^x)\tan(e^x)$ **(h)** $\dfrac{-\csc^2(2\sqrt{x})}{\sqrt{x}}$

6 (a) $\dfrac{dx}{dy} = -\sin y$

  (b) $\dfrac{dy}{dx} = \dfrac{1}{-\sin y} = \dfrac{1}{-\sqrt{1-\cos^2 y}} = -\dfrac{1}{\sqrt{1-x^2}}$

7 (a) $\dfrac{dx}{dy} = \sec^2 y$

  (b) $\dfrac{dy}{dx} = \dfrac{1}{\sec^2 y} = \dfrac{1}{1+\tan^2 y} = \dfrac{1}{1+x^2}$

8 (a) $\dfrac{dx}{dy} = \sec y \tan y$

  (b) $\dfrac{dy}{dx} = \dfrac{1}{\sec y \tan y} = \dfrac{1}{\sec y\sqrt{\sec^2 y - 1}} = \dfrac{1}{x\sqrt{x^2-1}}$

9 $\dfrac{dx}{dy} = -3\sin 3y,\ \dfrac{dy}{dx} = -\dfrac{1}{3\sqrt{1-\cos^2 3y}} = -\dfrac{1}{3\sqrt{1-x^2}}$

10 $\dfrac{dy}{dx} = -\sec x \tan x + \sec^2 x$

  $= \dfrac{\sin x + 1}{\cos^2 x}$

  $= \dfrac{1+\sin x}{1-\sin^2 x}$

  $= \dfrac{1}{1-\sin x}$

H Selecting methods (p 128)

H1 (a) Chain rule **(b)** Chain rule

  (c) Quotient rule **(d)** Product rule

H2 (a) $\dfrac{du}{dx} = 2e^{2x},\ \dfrac{dv}{dx} = \cos x$

  (b) $\dfrac{dy}{dx} = e^{2x}\cos x + 2e^{2x}\sin x$

H3 $\dfrac{du}{dx} = e^x,\ \dfrac{dv}{dx} = 4\cos 4x,\ \dfrac{dy}{dx} = \dfrac{e^x(\sin 4x - 4\cos 4x)}{\sin^2 4x}$

H4 $\dfrac{du}{dx} = 2e^{2x},\ \dfrac{dv}{dx} = -3\sin 3x,$

  $\dfrac{dy}{dx} = e^{2x}(2\cos 3x - 3\sin 3x)$

H5 $2xe^x + x^2 e^x + \dfrac{3}{x}$

H6 (a) $2x\cos(x^2+1)$ **(b)** $(x^2+1)\cos x + 2x\sin x$

Exercise H (p 130)

1 (a) $x^2 e^x + 2xe^x + 2x$ **(b)** $\dfrac{2x}{x^2-2}$

  (c) $-\tan x$ **(d)** $\dfrac{\sin x}{x} + \ln x \cos x$

  (e) $\dfrac{1}{x}\cos(\ln x)$ **(f)** $\dfrac{e^x(5x-4)}{(5x+1)^2}$

  (g) $\dfrac{-e^{-x}(2x+3)}{(2x+1)^2}$ **(h)** $\dfrac{e^{-2x}(2x-1)}{(1-x)^2}$

2 (a) $e^x \sin 3x + 3e^x \cos 3x$

  (b) $-e^{-x}(\cos 3x + 3\sin 3x)$

  (c) $e^{-2x}(\sec^2 x - 2\tan x)$

  (d) $3\cos 3x \cos 2x - 2\sin 3x \sin 2x$

  (e) $-3\sec^2(2-3x)$

  (f) $\dfrac{2\ln x + \dfrac{1}{x} - 2}{(1-2x)^2}$

3 $\dfrac{dy}{dx} = \dfrac{2x\cos 2x - \sin 2x}{x^2} - \dfrac{-\pi - 0}{\left(\dfrac{\pi}{2}\right)^2} = \dfrac{-4}{\pi}$

4 (a) 2 **(b)** 1 **(c)** -1

Mixed questions (p 131)

1 $\frac{1}{2}$

2 (a) $3e^{3x-2}$ **(b)** $\dfrac{3}{3x-2}$

  (c) $3\ln x + 3 - \dfrac{2}{x}$ **(d)** $12(3x-1)^3$

3 2.5

4 $\dfrac{dy}{dx} = 2x - 5,\ \dfrac{dy}{dx} = \dfrac{5-2x}{(x^2-5x)^2}$ and $\dfrac{dy}{dx} = \dfrac{2x-5}{x^2-5x}$

  are all zero when $2x - 5 = 0$; hence $x = 2.5$

5 $\dfrac{dx}{dy} = 3\sec^2 3y$

  $\dfrac{dy}{dx} = \dfrac{1}{\dfrac{dx}{dy}} = \dfrac{1}{3\sec^2 3y} = \dfrac{1}{3(1+\tan^2 3y)} = \dfrac{1}{3(1+x^2)}$

6 Let $u = e^x$ and $v = \cos 3x$

$$\frac{dy}{dx} = u\frac{dv}{dx} + v\frac{du}{dx}$$
$$= e^x(-3\sin 3x) + (\cos 3x)e^x$$
$$= e^x(-3\sin 3x + \cos 3x)$$

Expand $R\cos(3x + \alpha)$:

$R\cos 3x\cos\alpha - R\sin 3x\sin\alpha$ which is equivalent to $-3\sin 3x + \cos 3x$ with $R\cos\alpha = 1$ and $R\sin\alpha = 3$

So $\dfrac{dy}{dx}$ can be expressed in the form $Re^x\cos(3x + \alpha)$.

$R = 3.16$, $\alpha = 1.25$

Test yourself (p 131)

1 (a) $80\,°C$

(b) $e^{-0.1t} > 0$ for any t, although in this context $t \geq 0$ anyway so $20 + 60e^{-0.1t} > 20$

(c)

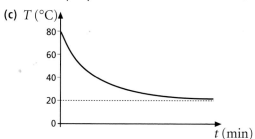

(d) 4.1   **(e)** $-6e^{-0.1t}$ **(f)** 38

2 (a) $3\cos(3x - 2)$ **(b)** $\dfrac{5e^{5\sqrt{x}}}{2\sqrt{x}}$

(c) $-\sec^2(4 - x)$ **(d)** $\dfrac{-2x}{7 - x^2} = \dfrac{2x}{x^2 - 7}$

3 (a) $e^{-x}(\cos x - \sin x)$ **(b)** $\dfrac{e^x(1 + \tan x)}{\cos x}$

(c) $18x(x^2 - 7)^8$ **(d)** $\dfrac{2x\sin x\cos x - x^2}{\sin^2 x}$

4 1

5 $\dfrac{dx}{dy} = \frac{1}{2}\cos\frac{1}{2}y$

$$\frac{dy}{dx} = \frac{1}{\dfrac{dx}{dy}} = \frac{1}{\frac{1}{2}\cos\frac{1}{2}y} = \frac{2}{\cos\frac{1}{2}y}$$

$\cos^2\frac{1}{2}y + \sin^2\frac{1}{2}y = 1$

so $\cos\frac{1}{2}y = \sqrt{1 - \sin^2\frac{1}{2}y} = \sqrt{1 - x^2}$,

so $\dfrac{dy}{dx} = \dfrac{2}{\sqrt{1 - x^2}}$

9 Numerical methods

A Locating roots (p 132)

A1 $f(2) = 3 + 2 - 2^2 = 1$

$f(3) = 3 + 3 - 3^2 = -3$

Exercise A (p 133)

1 $\ln 2 - 1 = -0.307$

$\ln 3 - 1 = 0.099$

The sign changes and $\ln x$ is continuous for $x > 0$ so there is a root between 2 and 3.

2 e^x and x^2 are continuous.

$e^{-1} - 3\times(-1)^2 = -2.632$

$e^0 - 3\times 0^2 = 1$

$e^1 - 3\times 1^2 = -0.282$

$e^3 - 3\times 3^2 = -6.914$

$e^4 - 3\times 4^2 = 6.598$

The sign changes between -1 and 0, between 0 and 1, and between 3 and 4, so there are roots in these intervals.

3 (a) $\sin(-3) - \frac{1}{-3} = 0.192$

$\sin(-2) - \frac{1}{-2} = -0.409$

$\sin 2 - \frac{1}{2} = 0.409$

$\sin 3 - \frac{1}{3} = -0.192$

The sign changes between -3 and -2, and between 2 and 3, and $\sin x$ and $\dfrac{1}{x}$ are both continuous in these intervals. So there are roots in these intervals.

(b) $\dfrac{1}{x}$ has a discontinuity at $x = 0$, so although there is a change of sign this does not indicate a root. Sketching indicates there is in fact no root.

4 $\cos x - \ln x = 0$

$\cos 1 - \ln 1 = 0.540$

$\cos 2 - \ln 2 = -1.109$

The sign changes and the function is continuous between 1 and 2 so there is a root in this interval.

5 $x^2 - \sqrt{x} - 2 = 0$

$1.8^2 - \sqrt{1.8} - 2 = -0.102$

$1.9^2 - \sqrt{1.9} - 2 = 0.232$

The sign changes and the function is continuous between 1.8 and 1.9 so there is at least one root in this interval.

6 **(a)** At A, $x^3 - 10 = \dfrac{1}{x}$

$\Rightarrow \quad x^4 - 10x = 1$

$\Rightarrow x^4 - 10x - 1 = 0$

(b) $2^4 - 10 \times 2 - 1 = -5$

$3^4 - 10 \times 3 - 1 = 50$

The sign changes between $x = 2$ and $x = 3$ and $x^4 - 10x - 1$ is continuous in this interval so there is a root in this interval. Hence the x-coordinate of A lies between 2 and 3.

7 **(a)**

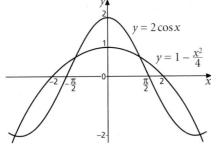

(b) 4

(c) The equation can be written as

$2\cos x + \dfrac{x^2}{4} - 1 = 0$

$2\cos(-1.3) + \dfrac{(-1.3)^2}{4} - 1 = -0.0425\ldots$

$2\cos(-1.2) + \dfrac{(-1.2)^2}{4} - 1 = 0.0847\ldots$

The sign changes and the function is continuous so there is a root between -1.3 and -1.2.

8 No, it is possible that over the interval $f(x)$ falls below 0 then returns to a positive value, giving two roots in the interval. (In general, there must be no roots or an even number of roots.)

9 Rearranging, $x^4 - 2x^3 - 4 = 0$

$(-1.089\,85)^4 - 2 \times (-1.089\,85)^3 - 4 = -0.000\,206$

$(-1.089\,95)^4 - 2 \times (-1.089\,95)^3 - 4 = 0.001\,024$

The sign changes and the function is continuous, so there is a root α in this interval.

To 5 s.f., $\alpha = -1.0899$

10 The equation can be written as $f(x) = 0$ where $f(x) = e^{-x} - x^3$.

When $x = 0.7725$, $f(x) = 0.000\,86\ldots$

When $x = 0.7735$, $f(x) = -0.001\,39\ldots$

The sign changes and $f(x)$ is continuous, so there is a root between 0.7725 and 0.7735.

So the root is 0.773 to 3 s.f.

11 $e^{2(0.567\,05)} - \dfrac{1}{(0.567\,05)^2} = -0.0016\ldots$

$e^{2(0.567\,15)} - \dfrac{1}{(0.567\,15)^2} = 0.0001\ldots$

The sign changes and the function is continuous.

So α lies between 0.567 05 and 0.567 15.

So $\alpha = 0.5671$ to 4 s.f.

12 $\sqrt{1.9965} + \cos 1.9965 - 1 = 0.00001\ldots$

$\sqrt{1.9975} + \cos 1.9975 - 1 = -0.00054\ldots$

The sign changes and the function is continuous in this interval.

So there is a root in the interval (1.9965, 1.9975).

So you can state the root to 3 d.p. (or 4 s.f.) as 1.997.

B Staircase and cobweb diagrams (p 134)

B1 **(a)**

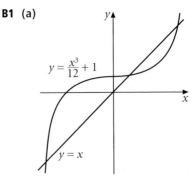

(b) 3

(c) This may be confirmed on the graph plotter or by testing for a root between 1 and 2 in $x^3 - 12x + 12 = 0$:

$1^3 - 12 \times 1 + 12 = 1$

$2^3 - 12 \times 2 + 12 = -4$

So there is a root between $x = 1$ and $x = 2$.

B2 Confirmation of the values shown

B3 The values are 1.22..., 1.15..., 1.12..., converging on 1.1157... after 10 iterations.

B4 Substituting $x = 1.1157$ into the cubic gives 0.000 408, which is close to zero.

B5 $x_1 = 1.083..., x_2 = 1.105..., x_3 = 1.112...,$ converging to 1.1157 after 7 iterations

B6 $e^{-0.2} - 0.2 = 0.619$

$e^{-1} - 1 = -0.632$

The sign changes and e^{-x} is continuous so there is a solution between $x = 0.2$ and $x = 1$.

B7 $x_1 = 0.81..., x_2 = 0.44..., x_3 = 0.64..., x_4 = 0.52...$

The values zigzag up and down but do eventually approach the limit 0.567...

B8 (a)

(i)
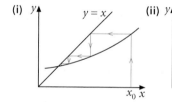

(ii)
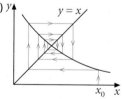

(iii)
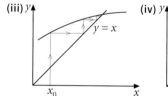

(iv)

(b) The curves with a positive gradient give a staircase. The curves with a negative gradient give a cobweb.

Exercise B (p 138)

1 (a) $x_1 = 1.4816, x_2 = 1.5000, x_3 = 1.5041$

(b) $x_1 = 1.5811, x_2 = 1.6066, x_3 = 1.6145$

(c) $x_1 = 1.9844, x_2 = 1.0233, x_3 = 1.8661$

(d) $x_1 = 0.3730, x_2 = 0.3630, x_3 = 0.3594$

2 (a) Q   **(b)** R   **(c)** P

3 (a) (i) $x_1 = 0.825, x_2 = 0.678, x_3 = 0.779$
limit = 0.739

(ii) $x - \cos x = 0$ (or an equivalent equation)

(b) (i) $x_1 = 1.667, x_2 = 1.600, x_3 = 1.625$
limit = 1.618

(ii) $x^2 - x - 1 = 0$ (or an equivalent)

(c) (i) $x_1 = 2.667, x_2 = 2.117, x_3 = 1.446$
limit = 0.458

(ii) $3x - 2^x = 0$ (or an equivalent)

(d) (i) $x_1 = 0.495, x_2 = 0.377, x_3 = 0.429$
limit = 0.414

(ii) $x^2 + 2x - 1 = 0$ (or an equivalent)

4 (a) $f(0) = -1, f(1) = 2$
The sign changes and $f(x)$ is continuous so there is a root between 0 and 1.

(b) $x^3 + 2x - 1 = 0$

$\Rightarrow \quad 2x = 1 - x^3$

$\Rightarrow \quad x = \dfrac{1 - x^3}{2}$

(c) $x_1 = 0.5000, x_2 = 0.4375, x_3 = 0.4581$

(d) $f(0.453\,35) = -0.000\,12...$
$f(0.453\,45) = 0.000\,13...$
The sign changes and $f(x)$ is continuous, so α lies between 0.453 35 and 0.453 45.
So α is 0.4534 to 4 d.p.

5 In each case if x is the limit $x_n = x_{n+1} = x$. Hence:

(a) $x = \dfrac{1 + 2x^2}{5}$

$\Rightarrow \quad 5x = 1 + 2x^2$

$\Rightarrow \quad 2x^2 - 5x + 1 = 0$

(b) $x = \dfrac{1}{5 - 2x}$

$\Rightarrow \quad (5 - 2x)x = 1$

$\Rightarrow \quad 5x - 2x^2 = 1$

$\Rightarrow \quad 2x^2 - 5x + 1 = 0$

(c) $x = \sqrt{\dfrac{5x - 1}{2}}$

$\Rightarrow \quad x^2 = \dfrac{5x - 1}{2}$

$\Rightarrow \quad 2x^2 = 5x - 1$

$\Rightarrow \quad 2x^2 - 5x + 1 = 0$

(d)
$$x = \tfrac{1}{2}\left(5 - \frac{1}{x}\right)$$
$$\Rightarrow \quad 2x = 5 - \frac{1}{x}$$
$$\Rightarrow \quad 2x^2 = 5x - 1$$
$$\Rightarrow \quad 2x^2 - 5x + 1 = 0$$

6 (a) $x_1 = 0.822\,071$, $x_2 = 0.818\,738$, $x_3 = 0.819\,238$

(b) At the limit, $x_n = x_{n+1} = k$ so $k = \dfrac{1}{\sqrt[3]{k+1}}$,

which can be rearranged to give $k^4 + k^3 - 1 = 0$.

7 $p = 3$, $q = 5$

8

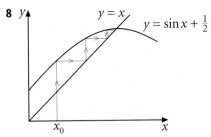

9 (a) 2

(b)
$$x^3 = 10$$
$$\Rightarrow \quad x^2 = \frac{10}{x} \quad (x \neq 0)$$

So $\quad x = \sqrt{\dfrac{10}{x}}$ (taking the positive square root)

The formula is $x_{n+1} = \sqrt{\dfrac{10}{x_n}}$.

$$x = 2.154$$

(c)
$$x^3 = 10$$
$$\Rightarrow \quad x^4 = 10x$$
So $x^2 = \sqrt{10x}$ (taking the positive square root)
So $\quad x = \sqrt{\sqrt{10x}}$ (again the positive square root)

The formula is $x_{n+1} = \sqrt{\sqrt{10x_n}}$.

(d) The second one is quicker.

10 (a) For example:

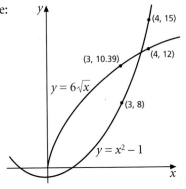

(b)
$$x^2 - 1 = 6\sqrt{x}$$
$$\Rightarrow \quad x^2 = 6\sqrt{x} + 1$$
So $\quad x = \sqrt{6\sqrt{x}+1}$
(taking the positive square root)

(c) 3.495

11 (a) $x^3 + x^2 - 3x - 1 = 0$
$$\Rightarrow \quad x^2(x+1) = 3x + 1$$
$$\Rightarrow \quad x^2 = \frac{3x+1}{x+1}$$
$$x = \sqrt{\frac{3x+1}{x+1}}$$

(b) $x_1 = 1.5811$, $x_2 = 1.4917$, $x_3 = 1.4823$

(c) Either of the following:

When $x_0 = -1$, $x + 1 = 0$ so $\dfrac{3x+1}{x+1}$ is not defined.

For $-1 < x_0 < -\tfrac{1}{3}$, $\dfrac{3x+1}{x+1}$ is negative so there is no real square root of it.

12 (a) (i) $x_1 = 7$, $x_2 = 342$, $x_3 = 40\,001\,687$,
$x_4 \approx 6.4 \times 10^{22}$

(ii) $x_1 = 6.25$, $x_2 = 0.0256$, $x_3 \approx 1526$,
$x_4 \approx 4.29 \times 10^{-7}$

(b) In neither case does the sequence converge. In (i) the values become ever more rapidly larger. In (ii) the outputs oscillate between ever larger and ever smaller values.

(c) (i)

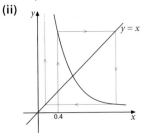

(ii)

(d) In both cases the problem arises where the curved graph is too steep, namely $|\text{gradient}| > 1$.

Mixed questions (p 141)

1 (a) At A, $x^3 + 1 = 2e^{-x}$

$\Rightarrow x^3 - 2e^{-x} + 1 = 0$ as required

(b) $0.5^3 - 2e^{-0.5} + 1 = -0.088...$

$0.6^3 - 2e^{-0.6} + 1 = 0.118...$

$x^3 - 2e^{-x} + 1$ is continuous and the sign changes so there is a root between 0.5 and 0.6.

2 (a)

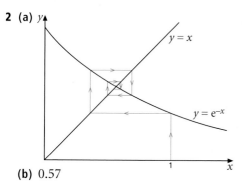

(b) 0.57

3 (a) 2.853

(b) $f(2.8525) = -0.000\,69...$

$f(2.8535) = 0.000\,65...$

The function is continuous and the sign changes. So the solution lies between 2.8525 and 2.8535. So the 4 s.f. answer of 2.853 is justified.

4 (a) $2^3 - 12 = -4$

$4^3 - 12 = 52$

The sign changes and the function is continuous, so there is a root in the interval.

(b) (i) $x = \dfrac{3x}{4} + \dfrac{3}{x^2}$

$\Rightarrow \quad x^3 = \dfrac{3x^3}{4} + 3$

$\Rightarrow \quad 4x^3 = 3x^3 + 12$

$\Rightarrow x^3 - 12 = 0$

(ii) $x_1 = 3.19, x_2 = 2.69, x_3 = 2.43$

(iii)

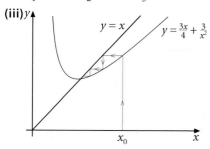

(iv) $\sqrt[3]{12}$

5 (a) $f(1.5) = -2.75$

$f(2) = 1$

The sign changes and $f(x)$ is continuous, so there is a root between 1.5 and 2.

(b) $x^3 - \dfrac{x^2}{2} - 5 = 0$

Dividing both sides by x $(x \neq 0)$,

$x^2 - \dfrac{x}{2} - \dfrac{5}{x} = 0$

$\Rightarrow \qquad x^2 = \dfrac{x}{2} + \dfrac{5}{x}$

Taking the square root of both sides $(x > 0)$,

$$x = \sqrt{\dfrac{x}{2} + \dfrac{5}{x}} \quad \text{as required}$$

(c) $x_1 = 2.021, x_2 = 1.867, x_3 = 1.900$

(d) $f(1.893\,85) = -0.000\,72...$

$f(1.893\,95) = 0.000\,16...$

The sign changes and the function is continuous over this interval.

So α lies between 1.893 85 and 1.893 95.

So $\alpha = 1.8939$ to 4 d.p.

6 $a = 2, b = 3, c = 4$

The required root $= 1.5747$ to 4 d.p.

Test yourself (p 143)

1 The equation can be rearranged as

$\sqrt{x} - 2\ln x = 0$

$\sqrt{2} - 2\ln 2 = 0.0279...$

$\sqrt{2.1} - 2\ln 2.1 = -0.0347...$

The sign changes and the function is continuous so there is a root in the interval.

2 (a)

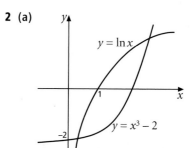

(b) 2

(c) The equation may be written as
$x^3 - \ln x - 2 = 0$
$1.3^3 - \ln(1.3) - 2 = -0.065\ldots$
$1.4^3 - \ln(1.4) - 2 = 0.407\ldots$
The sign changes and the function is continuous, so there is a root between 1.3 and 1.4.

3 (a) $x_1 = 2.646$, $x_2 = 2.940$, $x_3 = 2.990$

(b) (i) At the limit, $K = x_n = x_{n+1}$
substituting into the iterative formula,
$$K = \sqrt{K + 6}$$
Squaring both sides,
$$K^2 = K + 6$$
$$\Rightarrow K^2 - K - 6 = 0$$

(ii) Factorising,
$(K + 2)(K - 3) = 0$
$\Rightarrow \qquad K = -2 \text{ or } 3$
$K = 3$ is the required limit as the iterative formula always produces a positive square root.

4 (a)

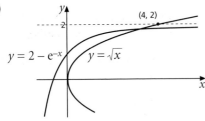

(b) $f(x) = 0$ is solved where the two graphs intersect. As $2 - e^{-x}$ is never greater than 2, the two graphs only intersect once, where shown. So $f(x) = 0$ has only one solution.

(c) $f(3) = -0.218\ldots$
$f(4) = 0.018\ldots$
The sign changes and the function is continuous over this interval, so the root lies between 3 and 4.

(d) $x_1 = 3.9270\ldots$, $x_2 = 3.9215\ldots$, $x_3 = 3.9211\ldots$,
$x_4 = 3.9211\ldots$
Hence the solution is 3.921 to 3 d.p.

5 $x = \sqrt{ax^3 - 3}$
Squaring both sides,
$$x^2 = ax^3 - 3$$
$$\Rightarrow ax^3 - x^2 - 3 = 0$$
Multiplying both sides by 4 to obtain the required coefficient of x and constant,
$$4ax^3 - 4x^2 - 12 = 0$$
Putting $a = \frac{1}{4}$ gives
$$x^3 - 4x^2 - 12 = 0 \quad \text{as required}$$

6 (a) $x^3 + x^2 - 4x - 1 = 0$
$\Rightarrow \qquad x^3 + x^2 = 4x + 1$
$\Rightarrow \qquad x^2(x + 1) = 4x + 1$
Dividing by $x + 1$ $(x \neq -1)$,
$$x^2 = \frac{4x + 1}{x + 1}$$
Taking the positive square root of both sides,
$$x = \sqrt{\frac{4x + 1}{x + 1}} \quad \text{as required}$$

(b) $x_2 = 1.58$, $x_3 = 1.68$, $x_4 = 1.70$

(c) $f(1.695) = -0.037\ldots$
$f(1.705) = 0.043\ldots$
The sign changes and the function is continuous in this interval, so α lies between 1.695 and 1.705.
Hence $\alpha = 1.70$ to 2 d.p.

(d) If $x_1 = -1$, the quotient inside the square root sign is not defined.
For $-1 < x_1 < -0.25$, the quotient inside the square root sign is negative so there is not a real square root.

10 Proof

A Introducing proof (p 144)

A1 With a 10-year-old child you might start by showing that an odd number of counters can be arranged in two equal rows with one counter left over. Then use a particular case such as 7 + 5, which can be represented by

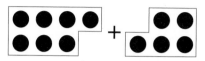

Since there are two odd numbers then the two 'left over' counters fit together so that there are two equal rows and hence an even number, which in this case is 12.

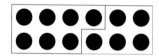

Since you will be able to produce pictures like this for any two odd numbers then you will always get an even number when you add two odd numbers.

In a mathematics examination, a briefer algebraic argument that does not appeal to pictures is more appropriate. It has the advantage that it shows that the statement is valid both for positive and for negative numbers.

Any two odd numbers can be written as $2k + 1$ and $2m + 1$ where k and m are integers.

The sum is
$$(2k + 1) + (2m + 1) = 2k + 2m + 2$$
$$= 2(k + m + 1)$$
Since $k + m + 1$ is an integer then $2(k + m + 1)$ is an even number and so the sum of any two odd numbers is always an even number.

A2 Some possible comments are as follows.

Response 1

After trying a few examples like this, it is difficult to believe that you will be able to find one where the identity will not hold true. However, no matter how far you go, this argument will not constitute a proof as it doesn't show that the identity is true for all values of x.

Response 2

This is a valid proof that shows the identity is true for all values of x. It relies on the distributive law that $a(b + c) = ab + ac$ for all real numbers. Whether or not you find this proof convincing will depend on your knowledge and experience of expanding brackets and simplifying algebraic expressions.

Response 3

This does not require as much knowledge of how to manipulate algebraic expressions and some may find the visual argument more convincing. An accompanying commentary that explains what is going on would possibly make things clearer.

Its main disadvantage is that it shows that the identity is valid for positive numbers greater than 1 whereas response 2 shows the identity is valid for all values of x.

B Disproof by counterexample

Exercise B (p 145)

1 There is only one possible counterexample and that is 2, the only even prime number.

2 Almost any value of θ gives a counterexample. The simplest is $\theta = 0$ as
$$\frac{\cos(3 \times 0)}{3} = \frac{\cos 0}{3} = \frac{1}{3} \text{ but } \cos 0 = 1.$$

3 A counterexample is given by 1 and 5, whose product is 5 which is not a multiple of 3.

4 The simplest counterexample is with $\theta = 0$ as $\cos(-0) = \cos 0 = 1$ but $-\cos 0 = -1$.

5 A counterexample is given by $\theta = \frac{\pi}{2}$ as
$$\sin\left(2 \times \frac{\pi}{2}\right) = \sin \pi = 0 \text{ but } 2 \sin\left(\frac{\pi}{2}\right) = 2.$$

6 A counterexample is with $x = \frac{1}{2}$ as $x^2 = \left(\frac{1}{2}\right)^2 = \frac{1}{4}$ and $\frac{1}{4}$ is smaller than $\frac{1}{2}$.

7 Almost any pair of values gives a counterexample. The simplest is $a = 1$, $b = 1$ as
$$(1 + 1)^2 = 2^2 = 4 \text{ but } 1^2 + 1^2 = 1 + 1 = 2.$$

8 The simplest counterexample is with $x = 0$ as
$$\text{cosec } 0 = \frac{1}{\sin 0} = \frac{1}{0} \text{ and this is not defined.}$$

9 A counterexample is given by the irrational numbers $\sqrt{2}$ and $\sqrt{8}$ as $\sqrt{2} \times \sqrt{8} = \sqrt{16} = 4$ which is rational.

10 Almost any pair of values gives a counterexample.
One pair is $A = \dfrac{\pi}{2}$, $B = \dfrac{\pi}{2}$ as

$\sin \dfrac{\pi}{2} + \sin \dfrac{\pi}{2} = 1 + 1 = 2$

but $\sin\left(\dfrac{\pi}{2} + \dfrac{\pi}{2}\right) = \sin \pi = 0$.

11 Almost any pair of values gives a counterexample.
One pair is $x = 9$, $y = 16$ as
$\sqrt{9+16} = \sqrt{25} = 5$ but $\sqrt{9} + \sqrt{16} = 3 + 4 = 7$.

12 Any value of x gives a counterexample.
The simplest is $x = 0$ as

$\dfrac{1}{2\cos 0} = \dfrac{1}{2}$ but $2\sec 0 = \dfrac{2}{\cos 0} = 2$

13 Almost any value of x gives a counterexample.
The simplest is $x = 0$ as
$e^{(2 \times 0)} = e^0 = 1$ but $e^0 + e^2 = 1 + e^2$.

14 Almost any pair of values gives a counterexample.
One pair is $A = 1$, $B = e$ as
$\ln(1 \times e) = \ln e = 1$ but $\ln 1 \times \ln e = 0 \times 1 = 0$.

15 A counterexample is given by $x = -4$ as
$|-4 + 2| = |-2| = 2$ which is less than 3 but
$|-4| = 4$ which is not less than 1.

16 A counterexample is given by $x = -4$, $y = 1$ as
$(-4)^2 = 16$ which is greater than $1^2 = 1$ but -4 is less than 1.

C Constructing a proof (p 146)

C1 Some experience of using the fact that an odd number can be written in the form $2k + 1$ for some integer k will be helpful (though not necessary). You would need to be able to square a simple linear expression ($2k + 1$ in this case) and factorise a simple expression ($4k^2 + 4k$). You also need to know that the product of an even number and a multiple of 4 will be a multiple of 8.

C2 The first conjecture is true (a counterexample for the second conjecture is that 3 is odd but $3^2 + 5 = 14$ which is not divisible by 3).
Proof:
If n is odd then there is some integer k such that $n = 2k + 1$.
So $n^2 + 3 = (2k + 1)^2 + 3$
$= 4k^2 + 4k + 1 + 3$
$= 4k^2 + 4k + 4$
$= 4(k^2 + k + 1)$
and since k is an integer, $k^2 + k + 1$ is an integer so $4(k^2 + k + 1)$ is a multiple of 4.
Hence, for all odd numbers n, $n^2 + 3$ is divisible by 4.

D Direct proof

Exercise D (p 148)

Each of these is just one possible proof.

1 Any odd number can be written as $2k + 1$ where k is some integer. Hence the next odd integer is $2k + 1 + 2 = 2k + 3$.
The sum is $(2k + 1) + (2k + 3) = 4k + 4$
$= 4(k + 1)$ which is a multiple of 4.

2 A trapezium $PQRS$ is shown below with the appropriate lengths labelled a, b and h.
The mid-point of QR is labelled O.

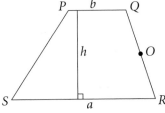

A rotation of $180°$ about O gives the shape $PS'P'S$.

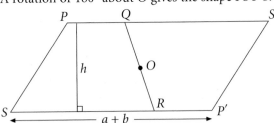

The length of SP' is $a + b$. The shape $PS'P'S$ is a parallelogram as its opposite angles are equal.
Hence its area is base $\times$ height $= h(a + b)$.

Now the area of trapezium $PQRS$ is half this area and so is $\frac{1}{2}h(a + b)$.

3 $\cos\theta$ and $\sin\theta$ can be defined as the x- and y-coordinates of the point P on the unit circle with centre $(0, 0)$, where θ is measured anticlockwise from the positive x-axis.

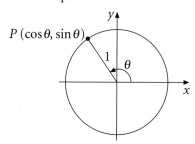

The equation of the circle is $x^2 + y^2 = 1$ so substituting for x and y gives $\cos^2\theta + \sin^2\theta = 1$.

4 The nth odd number is $2n - 1$.
So the sum S of the first n odd numbers can be written as
$S = 1 + 3 + 5 + 7 + \ldots + (2n - 3) + (2n - 1)$.
Reversing this sum gives
$S = (2n - 1) + (2n - 3) + \ldots + 7 + 5 + 3 + 1$.
Adding the first term from each series gives
$1 + (2n - 1) = 2n$.
Adding the second terms gives $3 + (2n - 3) = 2n$ and so on, giving n pairs of terms that each add to give $2n$. So $2S = 2n \times n = 2n^2$ which gives $S = n^2$ as required.

5 Four consecutive integers can be written as $k, k + 1, k + 2$ and $k + 3$ for some integer k.

The product of the last two integers is $(k + 2)(k + 3) = k^2 + 5k + 6$ and the product of the first two is $k(k + 1) = k^2 + k$.
The difference is $(k^2 + 5k + 6) - (k^2 + k) = 4k + 6$.

The sum of the four integers is
$k + k + 1 + k + 2 + k + 3 = 4k + 6$ which is the same as the difference, as required.

6 The values of both $\sin\theta$ and $\cos\theta$ are between -1 and 1 inclusive. Hence the lowest possible value of $\sin\theta\cos\theta$ is $1 \times -1 = -1$. So for all values of θ, the lowest possible value of $\sin\theta\cos\theta + 1$ is 0. So the graph of $y = \sin\theta\cos\theta + 1$ does not dip below the horizontal axis at any point and so the graph of $y = |\sin\theta\cos\theta + 1|$ is identical.

7 $\sin\theta\tan\theta \equiv \sin\theta \times \dfrac{\sin\theta}{\cos\theta}$

$\equiv \dfrac{\sin^2\theta}{\cos\theta}$

$\equiv \dfrac{1 - \cos^2\theta}{\cos\theta}$

$\equiv \dfrac{1}{\cos\theta} - \dfrac{\cos^2\theta}{\cos\theta}$

$\equiv \sec\theta - \cos\theta$　　as required

8 (a) Differentiating $\left(1 - \frac{1}{2}x^2\right)\cos x + x\sin x$ by using the product rule for each term gives
$\left(1 - \frac{1}{2}x^2\right)(-\sin x) - x\cos x + x\cos x + \sin x$
$= \frac{1}{2}x^2 \sin x$ as required.

(b) $\int x^2 \sin x \, dx = 2 \times \int \frac{1}{2}x^2 \sin x \, dx$

$= 2 \times \left(\left(1 - \frac{1}{2}x^2\right)\cos x + x\sin x\right) + c$

$= (2 - x^2)\cos x + 2x\sin x + c$

9 Four consecutive integers can be written as $k, k + 1, k + 2$ and $k + 3$ for some integer k.

The product of these integers is
$k(k + 1)(k + 2)(k + 3) = (k^2 + k)(k^2 + 5k + 6)$
$= k^4 + 5k^3 + 6k^2 + k^3 + 5k^2 + 6k$
$= k^4 + 6k^3 + 11k^2 + 6k$

Now the perfect square $(k^2 + 3k + 1)^2$ is $(k^2 + 3k + 1)(k^2 + 3k + 1)$
$= k^4 + 3k^3 + k^2 + 3k^3 + 9k^2 + 3k + k^2 + 3k + 1$
$= k^4 + 6k^3 + 11k^2 + 6k + 1$ which is one more than the expression above.

Hence the product of four consecutive integers is one less than a perfect square.

10 $\dfrac{\tan^2 A + \cos^2 A}{\sin A + \sec A} \equiv \dfrac{\sec^2 A - 1 + \cos^2 A}{\sin A + \sec A}$

$\equiv \dfrac{\sec^2 A - \left(1 - \cos^2 A\right)}{\sin A + \sec A}$

$\equiv \dfrac{\sec^2 A - \sin^2 A}{\sin A + \sec A}$

$\equiv \dfrac{(\sec A - \sin A)(\sec A + \sin A)}{\sin A + \sec A}$

$\equiv \sec A - \sin A$　　as required

11 $k^3 - k = k(k^2 - 1) = k(k + 1)(k - 1)$
$= (k - 1)k(k + 1)$ which is the product of three consecutive integers (as $k \in \mathbb{Z}$).

In any set of three consecutive integers, at least one is divisible by 2 and one is divisible by 3. Hence the product is divisible by 6 as required.

12 $9^n - 1 = 3^{2n} - 1 = (3^n - 1)(3^n + 1)$.

Now 3^n is odd for all n so both $3^n - 1$ and $3^n + 1$ are even.

$3^n - 1 + 2 = 3^n + 1$ so $3^n - 1$ and $3^n + 1$ are consecutive even numbers. In any pair of consecutive even numbers, one must be a multiple of 4. So the product $(3^n - 1)(3^n + 1)$ is the product of an even number and a multiple of 4. Hence the product is a multiple of 8 as required.

13 $p^2 - 1 = (p - 1)(p + 1)$

$p - 1$, p and $p + 1$ are three consecutive integers so one of them must be a multiple of 3. Now p is a prime greater than 3 so p is not a multiple of 3. Hence either $p - 1$ or $p + 1$ is a multiple of 3 and so the product $p^2 - 1$ is a multiple of 3.

Also, p is a prime greater than 2 so p must be odd. Hence there exists an integer k so that $p = 2k + 1$ and $p^2 - 1 = (p - 1)(p + 1)$ $= 2k(2k + 2) = 4k(k + 1)$.

Now, since k and $k + 1$ are consecutive integers, one of them must be even and so $k(k + 1)$ is even. So $4k(k + 1)$ is the product of 4 and an even number and so is a multiple of 8.

Since $p^2 - 1$ is a multiple of 3 and of 8 (and 3 and 8 have no common factor except 1) then $p^2 - 1$ is a multiple of $3 \times 8 = 24$.

E Proof by contradiction

Exercise E (p 150)

1 (a) Assume that $\sqrt[3]{2}$ is rational. Then there must exist integers p and q such that $\sqrt[3]{2} = \dfrac{p}{q}$ in its simplest form.

$\sqrt[3]{2} = \dfrac{p}{q} \implies p^3 = 2q^3$.

So p^3 is even and therefore p is even too.

If p is even then there exists an integer k such that $p = 2k$ and so $p^3 = (2k)^3 = 8k^3$.

Hence, $2q^3 = 8k^3$ which gives $q^3 = 4k^3$ and so q^3 is even and therefore q is even too.

But, if both p and q are even, then $\dfrac{p}{q}$ is not in its simplest form and our original assumption was false.

We conclude that $\sqrt[3]{2}$ is not rational.

(b) Assume that the equation $x^4 = 45x + 1$ does have at least one integer solution and call it k. Hence $k^4 = 45k + 1$.

Now k is either even or odd.

If k is even then k^4 is even and $45k + 1$ is odd so $k^4 \neq 45k + 1$.

If k is odd then k^4 is odd and $45k + 1$ is even so again $k^4 \neq 45k + 1$.

Hence our original assumption is false and the equation $x^4 = 45x + 1$ has no integer solutions.

(c) Assume that there is at least one pair of positive integers x, y such that $x^2 - y^2 = 10$.

Now $x^2 - y^2 \equiv (x - y)(x + y)$.

If x and y are both even, then both $(x - y)$ and $(x + y)$ are even so the product $(x - y)(x + y)$ is a multiple of 4.

If x and y are both odd, then both $(x - y)$ and $(x + y)$ are even so the product $(x - y)(x + y)$ is a multiple of 4.

If x is even and y is odd (or vice versa), then both $(x - y)$ and $(x + y)$ are odd so the product $(x - y)(x + y)$ is odd.

Each of these contradicts our original assumption that $x^2 - y^2 = 10$.

Hence no such integers exist.

2 Jo must have reasoned something like this. 'If I have the £50 note, then Raj would now be looking at it and he would say straight away that he had a £5 note. He isn't saying anything so I cannot have the £50 note. Hence I must have a £5 note.'

F Convincing but flawed

Exercise F (p 151)

1 As $a = b$, $a^2 - ab = a^2 - a^2 = 0$.
So the last step is equivalent to dividing by 0 which is impossible (for example, we know that $1 \times 0 = 2 \times 0$ but this does not imply that $1 = 2$) and leads to the contradiction.

2 $n + n$ represents the sum of two identical numbers which is indeed even but, if the numbers are different, then you need two different letters such as m and n to represent the two different numbers. You cannot conclude that $m + n$ is even.

3 The mistake here is on line 3.

The expression on the right-hand side should be $\pm\sqrt{1-\sin^2\theta}$ to include both the positive and negative square roots. This would give

$1 + \cos\pi = 1 \pm \sqrt{1-\sin^2\pi}$ on line 6 which does not lead to a contradiction as

$1 + \cos\pi = 1 - \sqrt{1-\sin^2\pi}$ is true.

4 The mistake here is to assume that, if a statement is true for many cases, then it will be true for all cases. It is clear, for example, that when $n = 41$, the expression is equivalent to 41^2 which is not prime.

5 The mistake here is to assume that, if the sides of a polygon are all equal, then it will have equal angles (clearly not true in the case of a rhombus that is not a square for example). It also, on first glance, looks like a regular octagon. However, we can use symmetry and trigonometry to show that $\angle A = \angle C = \angle E = \angle G \approx 126.9°$ and $\angle B = \angle D = \angle F = \angle H \approx 143.1°$.

6 The mistake here is to assume that a pattern will continue in the obvious way. Here 5 points gives 16 regions (following the doubling pattern) but 6 points gives 31 regions (not the expected 32).

The maximum number of regions is in fact $\frac{1}{24}(n^4 - 6n^3 + 23n^2 - 18n + 24)$.

(For those who would like to try proving this, a proof can be obtained using Euler's formula for connected networks that can be drawn on the surface of a sphere.)

7 The mistake here is to assume that the large shapes are in fact triangles. Consider the smaller triangle at the top of the first shape. The gradient of the hypotenuse is $\frac{3}{5} = 0.6$. However, the gradient of the hypotenuse of the larger triangle is $\frac{5}{8} = 0.625$. As the gradients are close the lines look as though they form the hypotenuse of a large right-angled triangle but they don't. The large shape is in fact a convex quadrilateral.

The second shape is not a triangle for the same reasons and is in fact a concave quadrilateral with a smaller area than the first quadrilateral.

The area of the black shape is the difference between their areas.

Test yourself (p 153)

1 The smallest counterexample is the even number 10 as it can be expressed as the sum of two primes in two different ways, $3 + 7$ and $5 + 5$.

2 (a) This statement is false. A counterexample is given by $n = 4$ as $f(4) = 4^2 + 4 + 1 = 21$ which is not prime.

(b) If n is odd, there is an integer k such that $n = 2k + 1$.

So $n^2 + n + 1 = (2k+1)^2 + (2k+1) + 1$
$= 4k^2 + 4k + 1 + 2k + 1 + 1$
$= 4k^2 + 6k + 3$
$= 2(2k^2 + 3k + 1) + 1,$

which is odd.

If n is even, there is an integer k such that $n = 2k$.

So $n^2 + n + 1 = (2k)^2 + 2k + 1$
$= 4k^2 + 2k + 1$
$= 2(2k^2 + k) = 1,$

which is odd.

Hence, $f(n)$ is always odd.

3 (a) The simplest pair of values that give a counterexample is $A = 0, B = 0$ as $\sec(0 + 0) = \sec 0 = 1$ but $\sec 0 + \sec 0 = 1 + 1 = 2$.

(b) $\tan\theta + \cot\theta \equiv \dfrac{\sin\theta}{\cos\theta} + \dfrac{\cos\theta}{\sin\theta}$

$\equiv \dfrac{\sin^2\theta + \cos^2\theta}{\cos\theta\sin\theta}$

$\equiv \dfrac{1}{\cos\theta\sin\theta}$

$\equiv \dfrac{2}{2\cos\theta\sin\theta}$

$\equiv \dfrac{2}{\sin 2\theta}$

$\equiv 2\csc 2\theta$ as required.

4 (a) Four consecutive integers can be written as $k, k + 1, k + 2$ and $k + 3$ for some integer k.

The product of the first and last integers is $k(k + 3) = k^2 + 3k$ and the product of the middle two is $(k + 1)(k + 2) = k^2 + 3k + 2$. The difference is $(k^2 + 3k + 2) - (k^2 + 3k) = 2$ as required.

(b) Three consecutive multiples of 4 can be written as $4k$, $4k + 4$ and $4k + 8$ for some integer k.

The sum is $4k + 4k + 4 + 4k + 8 = 12k + 12$ $= 12(k + 1)$ which is a multiple of 12.

(c) $\ln a^3 - \ln a^2 - \ln a \equiv 3\ln a - 2\ln a - \ln a$
$\equiv 0$ as required.

Index